REARING QUEEN
HONEY BEES

REARING QUEEN HONEY BEES

ROGER A. MORSE

WICWAS PRESS · CHESHIRE · CONNECTICUT

1994

Second edition
Copyright © 1994 Wicwas Press

First edition Copyright © 1979
First edition reprinted in 1982 and 1984

Cover design by Lawrence J. Connor

Library of Congress Cataloging-in-Publication Data
Morse, Roger A.
    Rearing queen honey bees / Roger A. Morse. -- 2nd ed.
       p. cm.
    Includes bibliographical references and index.
    ISBN 1-878075-05-5
    1. Bee culture--Queen rearing. I. Title
SF531.55.M67 1993
638'.145--dc20   93-41399
                                           CIP

TABLE OF CONTENTS

PREFACE

James Powers, a successful commercial migratory beekeeper who once operated 34,000 colonies of bees in five states, told me that "Queen rearing is too important a task to leave to someone else." He grew all of his own queens. On the other hand, Paul Englehart, a successful New York state beekeeper who once had 3,000 colonies, purchased the queens he used. However, he likewise was a cautious beekeeper. He told me that in the spring, just before it was time to start queen rearing, he went south and visited the beekeeper who grew his queens. Paul needed only time for a cup of coffee to ensure that the lawn was mowed, the shop clean, the hives properly painted, the queen cups ready, and that the beekeeper and his wife were on speaking terms. If everything was in order, Paul knew the queen breeder could concentrate on his job and deliver good queens.

Queen rearing requires attention to detail. Queens vary greatly in size and weight; the greater the weight, the more ovarioles[1] a queen has and the more eggs she will lay. The size of the queen is a direct result of how well she is fed and cared for during her growth and development, especially during the larval stage. Better queens are produced when nectar and pollen are available in great quantities. If a nectar and/or pollen flow stops abruptly, the beekeeper must step in and take measures to make certain quality queens are produced.

All this does not mean that one cannot buy queens. Still, the individual who rears his own queens - especially if he or she rears only a few - can select the best time of year, the time when both

---

[1] The late Professor J. E. Eckert of the University of California wrote about this in the June, 1934 issue of the *Journal of Economic Entomology*. He found that the total number of ovarioles in a queen's two ovaries could vary from 260 to 373. Eckert examined about 300 queens to obtain these figures.

nectar and pollen are abundant, to do so. This is a great advantage.

Queen rearing is a specialized area, but with patience and care it can be mastered. This book is designed for the beekeeper who would rear queens, whether a few or a thousand or more. The basic needs and equipment are very much the same.

CHAPTER 1

INTRODUCTION

There is nothing secret or mysterious about being successful in beekeeping. There are some very simple guidelines. Two of these are that good queens produce more offspring than poor ones, and queens less than a year old are much less likely to head colonies that may swarm than older queens are. Under ideal conditions, one wants the maximum number of bees in a hive during the honey flow; this means maximum brood production and the prevention of swarming, which divides a colony's strength.

One should grow enough queens to requeen every colony every year, provided, of course, that one is aiming at maximum production. And, since some unforeseen misfortune may result in losing a queen, it is good to have a few extra queens always available in nucleus colonies; these may be used to requeen in an emergency at any time of year.

I often have been criticized for placing too much emphasis on honey production. Some people keep a few hives of bees for fun only. I also have met those who believe that keeping (and watching) bees can have therapeutic value. I have no objection to this, but I do believe that a person who seeks to maximize production really gets the most, both in terms of profit and enjoyment, from his or her vocation or avocation.

RACES OF BEES

About 26 races of honey bees are native to Europe, Africa, and the Near East. None of these races is pure today because of the

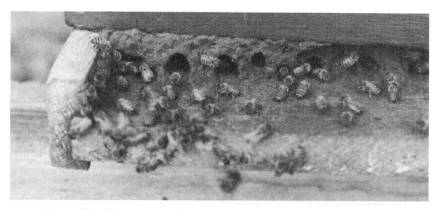

1. Propolis, the gums and resins collected by bees, fills the space between the bottom board and the hive body. The bees have left spaces about three-eights to one-half inch in diameter through which they may enter the hive. Plugging an entrance in this manner allows the colony to protect itself with ease, but the sticky propolis is a nuisance to beekeepers. Propolizing an entrance is a trait of Caucasian bees.

tremendous movement of honey bees around the Earth since the last century. However, many beekeepers and researchers have bees that are predominantly of one race.

Three races have been favorites during the past 50 years: Italian, Carniolan, and Caucasian. Until the past few years the Italian bees were preferred by beekeepers, and the bee journals frequently ran ads for "Italian" or "Golden Italian" queens. The Italian bees have several good characteristics: they are moderately resistant to European foulbrood, use little propolis, are relatively gentle, lend themselves to manipulation and swarm control techniques, and are good producers. However, Italian bees are susceptible to the American strain of chalkbrood and tracheal mites - especially the latter - and thus have gone out of favor in many areas. While the Italian bees are capable of surviving a chalkbrood attack, they cannot withstand tracheal mites.

Caucasian and Carniolan bees appear to be quite resistant to tracheal mites. Losses among these races have been severe, but the most susceptible of these bees apparently have died in the past few years. Caucasian and Carniolan bees, especially the latter, use too much propolis, but hopefully we can select this characteristic out of the stock we are now using. Both races are less aggressive than most other bees and have found much favor for this reason. They are reasonably good producers.

During the 1990s it is more important to select bees that are free of tracheal mites and chalkbrood than to be concerned about other criteria. Hopefully other considerations, such as varroa mite resistance, will become more important in the near future.

Where can one grow queens?

One may grow a few queens almost anywhere. I say "almost" because queen rearing does require that there be a large number of drones available for successful mating to take place. There has been little research on the question but probably one must have a minimum of 3,000 to 5,000 drones in the nearby area. One study on this question was done over 70 years ago on an island in Lake Ontario.[1] Still, in my conversations with commercial queen breeders, they have confirmed that without a great number of drones their queens are not fully mated. This must be taken into account by persons who grow queens in mountainous regions or on islands.[2]

Growing queens in cities may prove difficult, too, because queen rearing requires that the beekeeper inspect his or her colonies at precise times. Most city beekeepers are careful to manipulate their colonies only when the weather is warm and the sun is shining, because doing so at other times may arouse the

---

[1] This question was discussed by Dr. E. Oertel in an article entitled "Queen mating experiments fifty years ago" in the October, 1971 issue of *Gleanings in Bee Culture*.

[2] Dr. R. H. Anderson reported at the 1976 Apimondia conference on *African Bees, Taxonomy, Biology and Economic Use* that when he established a single mating apiary on a remote island off the South African coast, the queens "seemed reluctant to mate" (at least normal mating was not taking place). When he established a second apiary on the same island, mating was normal. He indicates that there is some deterrent which prevents queens from mating with drones from their own apiary. Anderson was working with *adonsoni* bees native to southern Africa; these may behave differently than European bees.

13

2. During good weather, and especially when nectar is available in quantity, it is possible to inspect a colony without wearing a veil. One should always have a smoker available to aid in calming the bees in case of error, such as dropping a comb.

bees unduly and perhaps cause them to sting neighbors. When one has ripe queen cells (cells from which virgin queens are about to emerge), they must be removed from the cell finishing colonies, separated, and put into mating nuclei on time, otherwise the virgins may emerge and kill one another. This often forces the beekeeper to work in inclement weather.

If one uses the grafting technique—that is, the transfer of young larvae from a brood comb to queen cells—it is almost necessary to have a small room or building in which to work. A good light is helpful. I once visited a beekeeper whose wife grafted in the back of a small van; she had all the necessary equipment including a good light. Even when one uses other methods to rear a few queens, it is convenient to have a good workbench. All this means is that the best place to grow queens may be in one's backyard; a shop or the kitchen may be used for the more delicate operations.

The last requirement, which will be emphasized over and over, is that there be an abundance of pollen- and nectar-producing plants during the time the queens are raised. This may require that queen rearing be restricted to a particular time of the year. Researchers have been able to grow queens all year using pollen substitutes and supplements; however, insofar as commercial or hobby beekeeping is concerned, that is not practical.

15

3. *This is a natural swarm clustered on a small tree. When a swarm emerges from its hive, it clusters in a temporary location such as this while the scout bees confirm the suitability of the new home site. Natural swarms which cluster close to the ground are easy to capture and hive. The bees in a swarm are engorged and their wax glands well developed. Such bees may be used in a variety of ways, including drawing foundation, for comb honey production or in starter colonies (see Chapter 5).*

CHAPTER 2

LIFE HISTORY RELATING TO QUEEN PRODUCTION

A queen bee is an unusual animal. She may live one, two, or as long as five years. Worker bees live for a much shorter period of time, usually only about six weeks in the summer and several months in the winter. Queens and workers are both females. To grow queens one starts with one-day-old larvae destined to become worker bees, and changes their environment so that hive bees will feed them differently and cause them to become queens. To grow queens successfully it is only necessary to understand the biology and life history of honey bees. To produce queens, we create conditions within the hive under which the best natural queens are produced.

LIFE HISTORY OF QUEENS, WORKERS, AND DRONES

Queens, workers, and drones all arise from the same egg. Eggs that develop into drones are not fertilized. Queens and workers both arise from fertilized eggs. Whether a queen or a worker develops depends upon the kind of food received during larval life. However, even the larval food is the same until the end of the second day. Nurse bees feed young larvae royal jelly, a secretion from glands in their heads. Sometime during the sec-

Table 1. Development time (in days) for queens, workers, and drones.

Stage	Drone	Worker	Queen
Egg	3	3	3
Larva	6.5	6	5.5
Pupa	14.5	12	7.5
Total	24	21	16

ond day the diet of those destined to become workers is slowly changed; honey and pollen are added and the diet is said to be more crude.

Larvae destined to become queens are fed royal jelly throughout their larval life. The food makes the difference, though precisely how this happens is a question still being studied. What is important for the queen breeder to know is that during the first 24 to 36 hours of larval life, a queen or a worker may develop. Some people have been misled into thinking that two-day-old larvae are satisfactory for queen rearing, but this is not so. In the laboratory, by taking larvae of different ages and controlling the food intake we can produce intermediates, creatures in between queens and workers, including many that are very queen-like. From a practical point of view it is not near-queens, but fully developed ones that we want to head our colonies.

Continued feeding of royal jelly not only causes differences in the adult produced, but also speeds up development. We presume it is the richer food that causes queens to grow faster. Thus, while worker bees develop in 21[1] days, only 16 days are required for queen development. This shortened development

---

[1]Some variation in development time occurs with some reports showing workers developing in as few as 19 or 20 days; cold weather may delay development for several days, too.

It is also correct that some variation may occur in the development of queens. Dr. R. D. Fell has data to show that bees may rear queens from larvae older than is reported in the literature; however, these are probably not fully developed queens, and while the colony may survive with these present, we presume such queens lay fewer eggs.

time for queens is important in natural requeening should a colony accidentally lose its queen, for it reduces the time required to grow a new queen. Table 1 outlines the normal development time for the hive occupants.

STAGES AND VARIATIONS IN QUEEN CELL DEVELOPMENT

Queens are reared under one of three conditions: when a colony has lost its queen, when it is preparing to swarm, and when the queen is replaced because she is old and no longer producing the chemicals by which she is recognized. This last act is called supersedure.

There are two types of queen cells: emergency and natural. The stages in natural queen cell development are as follows: queen cups, queen cups with eggs, developing queen cells with larvae, capped queen cells, and sometimes capped queen cells with tips removed.

Emergency queen cells are those that are built after the

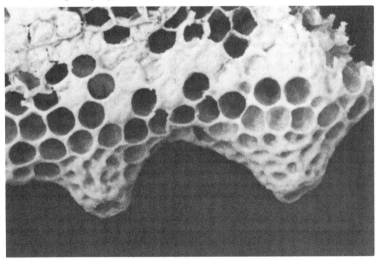

4. Here are two natural queen cups hanging from the bottom of this comb. Bees build queen cups several weeks before the queen may lay in them. The construction of queen cups is the first sign that a colony is crowded and may swarm.

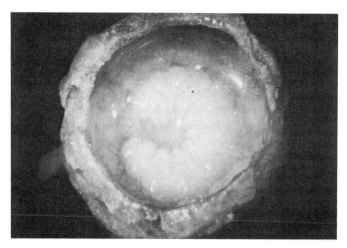

5. *A two-day-old larva floats in a sea of royal jelly in the queen cell. Worker bees do not feed larvae directly but keep an abundance of royal jelly in the cells. The larvae feed by turning around and around in their cells.*

sudden loss of the queen. They are cells built around worker larvae, which the workers have suddenly determined will become queens because of the emergency situation.

Natural queen cells are more commonly built on the bottom edge of the brood nest, but a small number may be found in the middle of the brood area. Emergency cells are all within the brood nest, since they are constructed around larvae hatched from eggs laid before the queen's loss; however, there is a tendency to build emergency cells near the bottom of the nest, too. In the case of supersedure, there are usually fewer cells constructed than during swarm preparations or emergency situations. Although the number may overlap, by examining the number and position of the queen cells the beekeeper has good clues as to why they are being constructed.

Queen cups: Natural queen cups are constructed in colonies several weeks before swarming may occur. Their construction is the first visible sign the beekeeper has that a colony may swarm. When one is rearing queens, human-made cups of beeswax or plastic cups can be provided and will be accepted by the bees.

Queen cups with eggs: The deposition of eggs in cups by a queen and the whitening of cup edges by the addition of new wax is the second visible step in natural swarming and queen production. When eggs are deposited in cups by a queen it is further evidence that a colony is crowded and may swarm; however, at this stage the process is still reversible and the beekeeper may prevent swarming if he or she takes the right steps.

Developing queen cells with larvae: Shortly after the egg hatches in a queen cup, worker bees deposit large quantities of royal jelly in the cell. The developing larva is not fed directly but floats in a sea of food, ever turning around and around in the cell as she feeds. Worker bees add wax to the cell edge, making it long enough to accommodate the larva when it stretches to its full length.

 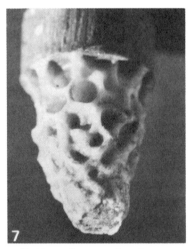

6. This is a mature, capped queen cell. Note the wooden, human-made cell base. 7. This is a mature queen cell with the cell tip removed. Worker bees may remove the tip anytime after the cell is capped and before the queen emerges. When the tip is removed, the cocoon is exposed; some have said this may make it easier for the queen to cut her way out of her cell, but this is questionable. We believe that since the cocoon does not touch the cell tip, the tip is therefore not contaminated with secretions from the cocoon, allowing the bees to remove it. The chemical nature of these secretions is under investigation.

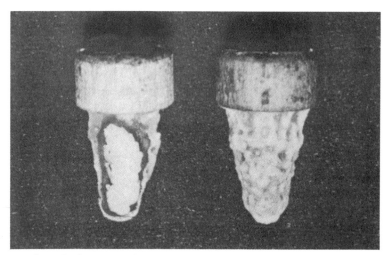

8. *When she has completed feeding and cocoon spinning, the queen larva stretches lengthwise in her cell and transforms into a pupa. The side of the cell on the left has been cut away to expose the developing queen. This is a delicate stage in the life of the queen; if the cell is jarred or shaken, the pupa will be killed.*

Capped queen cells: The precise time the queen cell is capped can vary by several hours, but it generally takes place around the end of the fifth day after the egg hatches. The larva continues to feed on the royal jelly, which is always present in extra supply; in fact, there is usually royal jelly left in the bottom of the queen cell after the larva has advanced to the pupal stage age and while the cocoon is being spun. Queen cocoons do not envelop the whole larva as those of workers and drones do; rather, the cocoon is constructed around the sides and tip of the cell.

Queen cell tip removal: In the case of the queen cell, the cocoon does not touch the tip of the cell. Queen cells are the only objects bees build that they allow to obstruct the bee space, or the walking space between and around the combs. We believe that, as the cocoon is spun, a chemical substance (pheromone) is deposited with the silk which inhibits the bees from removing what would otherwise be treated as something in their way. Since the cocoon does not touch the wax tip of the cell, that portion of the cell has no such message imbedded in the wax. Sometime after

22

the cocoon is spun, worker bees may remove the wax tip. Bee-keepers have long thought that the removal of this wax is to help the queen as she chews her way out of a cell. It is interesting that the wax tip is not always removed. We have observed that queens emerge, apparently without difficulty, from queen cells from which the tip has not been removed.

Still another concern of queen cell construction is the extent to which the cell exterior is chewed or mottled. Queen breeders prefer cells with rough exteriors, scalloped to a considerable degree by the bees. Queens reared in cells with smooth exteriors are said to receive less attention and probably have been fed less during larval life. I know of no data on this subject, but share the strong prejudice toward queen cells that are light in color over those that are dark. Light ones have been constructed of new wax; if worker bees are secreting new wax, one may conclude that they are well fed. Bees will use old wax—that is, move wax from one place on a comb to another—when it is needed. A dark queen cell built of old wax mean the bees were not secreting new wax and probably were not so well fed.

VIRGIN QUEEN EMERGENCE, MATURATION, AND MATING

A virgin queen emerges from her cell at or near the end of the sixteenth day after the egg has been laid. She chews her way out of her cell without assistance from any worker bees. She engorges on honey (and perhaps pollen), again without assistance, or at least we presume. However, shortly after emergence the virgin queen learns to solicit food and thereafter no longer feeds herself. A newly emerged virgin queen apparently produces none of the chemical substances by which mature queens are recognized; these substances are acquired slowly, and usually by the third day of her adult life some workers are attracted to the young queen and are starting to lick and groom her.

Recently emerged virgin queens are generally more active than their worker sisters. They move about the brood nest, apparently searching for other virgin queens and/or queen cells. Virgin

9. This is a queen cell from which a queen has emerged. The young queen chews her way out of her cell by turning in a circular fashion within the cell. She is not assisted by the worker bees. A queen rarely chews away the whole cell tip and a flap remains hinged to the cell.

queens will fight with each other until only one survives. When capped queen cells are found, the virgin chews a hole in the side. She sometimes, but not always, stings the developing queen. As far as we can determine, her actions are not related to the age of the queen developing in the cell; that is, she does not necessarily sting the older ones and not the younger ones. However, researchers are not unanimous about this question and more research is needed. The act of opening and making a hole in the side of a queen cell apparently serves as a stimulus for worker bees to continue the cell's destruction and discard the contents. Virgin queens pay no noticeable attention to open queen cells containing developing larvae. However, the presence of a virgin queen in a hive seems to stimulate worker bees to cease feeding these larvae and destroy the cells and their contents, although this may not occur immediately. Since worker bees do not appear to recognize a virgin queen for the first few days of her life, it is not clear how they receive this message.

Within about three to five days of emerging from her cell, a virgin queen normally takes her first flight. Flights may last from

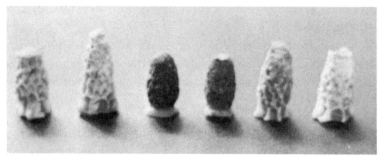

10. Here are six mature queen cells. The two in the center are much less well mottled and are also darker. This indicates they were not reared under ideal conditions. The four lighter colored, larger cells were made with freshly secreted wax indicating a honey flow was in progress (or that the colony was being heavily fed). The bees made the darker cells from wax removed from old comb. The fact that the dark cells were less well mottled indicates they were not so attractive to workers, probably because the developing queens produced less of the attractive secretion. Cells such as the two in the center should be discarded.

two to about 20 minutes. If the virgin queen's flight is of short duration, it is an orientation flight only. About half of the virgin queens mate on the first flight. It is probably important that a queen's flight time be short. Since they are relatively large insects, they are no doubt subject to predation by a large number of animals. See Chapter 12 for the special problem caused by dragonflies.

Virgin queens mate about 12 to 18 times in two or three flights over a period of two to four days. Mating is weather dependent and does not take place during bad weather. Before the mid-1950s it was thought that queen honey bees mated only once and rarely twice. Only recently has multiple mating and its implications been understood, and not all textbooks include this new information.

Virgins lose their ability to mate normally after they are about three weeks old. From a practical point of view it is best to destroy virgin queens that have been confined for three weeks because of inclement weather, and many beekeepers do so after two weeks. In no case should virgin queens be caged, mailed or otherwise confined. They should be allowed to fly and mate as soon as possible.

Queens start to lay eggs about two or three days after the last mating flight. Once egg laying has commenced, queens may be caged, used to requeen colonies or shipped without difficulty. As far as we are aware, no queen has ever been observed leaving the hive after she has started to lay eggs, except to accompany a swarm.

DRONE EMERGENCE, MATURATION, AND MATING

Drones emerge from their cells 24 days after the unfertilized eggs are laid. They engorge shortly after emergence but soon learn to solicit food from workers; once this process is learned the drones appear to be unable to feed themselves. In the fall when workers stop feeding drones, they still do not feed themselves and slowly starve.

Drones mature slowly. They are not capable of mating until they are eight to 12 days of age. Food, temperature and racial background all appear to affect the onset time of sexual maturity. Drones live on the edge of the brood nest. Workers apparently will not tolerate them in the middle of the brood nest and may drive them from it, though this has not been demonstrated.

Drone flights last longer than queen flights do; they may be as long as 55 minutes, although more often they are only about 30 minutes in duration. Drones do not stop to rest while searching for queens, but return to the hive when their food supply is nearly exhausted. No one has ever found a mature, normal drone resting in the field. Old, starved drones may be found on foliage (rarely on flowers) near hives in late fall, but these are the unfortunate creatures who have been cast out by workers preparing for winter.[2]

Because drone flights take longer, it is presumed that they fly greater distances than queens do. The data are not precise. It is logical that they do so, as this would tend to discourage the brother-sister matings which nature usually takes steps to discourage. If one is seeking to isolate a mating area, it is necessary that it be more than 10 miles from the nearest bees and source of drones. Dr. D. F. Peer, working with the Canada Department of Agriculture in the late 1950s, found a small number of queens were mated when they were separated by this distance from drones.[3] These queens probably did not mate as many times as queens are capable of mating.

Drones die in the act of mating. A dying (or dead) drone may sometimes be found after a mated pair has fallen to the ground.

---

[2]In 1991, Dr. M. Sasaki and his associates at Tamagawa University in Japan reported a great curiosity in the journal *Experimentia*. They observed drones of *Apis cerana*, the small Indian honey bee, visiting and aggressively mauling one species of orchid. Apparently, this orchid species produces a chemical that mimics the sex attractant in *Apis cerana*. Drones from this species do not visit other flowers, and drones from nearby European colonies did not visit this orchid species. Clearly, the sex attractant in the two honey bee species is different.
[3]This was reported by Dr. Peer in an article entitled, "Further studies of the mating range of the honey bee" in the March, 1957 issue of the *Canadian Entomologist*.

Usually, queens detach themselves from the drones after mating is complete, although they may sometimes return to the hive with the mating sign[4] still attached to them. When this occurs it is removed by worker bees. Drones that fail to find a queen to mate with may live for several weeks; like workers, their life span depends on how much flying they do.

Drone congregation areas

Mating in honey bees takes place in specific areas known as drone congregation areas. There are usually several such areas within flight range of any apiary. These areas remain the same year after year. While we have been able to locate some of these areas, we do not know how they come into being or why the choice of areas remains unchanged through the years.

Drones from colonies in many directions are attracted to congregation areas, which means it is difficult to control mating. It is only through saturating an area with drones, and by having one's neighbors interested in doing likewise with the same stock, that one may obtain some control over the outcome. It is very discouraging trying to mate Italian stock when one's neighbors have Caucasians or some other stock.

Individual matings are rapid, requiring less than a minute. Mating pairs meet high in the air, usually at a height of 20 to 50 or more feet. Worker bees usually fly within six to eight feet of the ground and thus the flight lanes of workers and those used by mating queens and drones are separate. Worker bees are antagonistic toward queens outside the hive and may attack and sting them, thus the importance of separation.

The beekeeper has little control over when and where mating will take place. In part, of course, this is weather dependent. Apiaries with proper windbreaks and a maximum of sunlight will encourage mating on days when there is inclement weather.

---

[4] Mating sign is the beekeeper's term for a portion of the male genitalia that the queen is sometimes unable to dislodge without assistance. For many years beekeepers incorrectly assumed that all queens retained and returned to the hive with such a sign. They thought that those who returned without sign had not mated.

Since it is important to have queens mate with a maximum number of drones within a short time period, it is helpful to select good apiary sites.

EGG LAYING

Queen breeders do not remove queens from their mating nucs until they have begun to produce eggs. They are not held long enough to judge their egg-laying pattern, only to determine that they are laying normal eggs in a normal position; in fact, a nuc box is really not big enough for one to judge a queen's brood pattern (brood patterns are discussed at length in Chapter 9). If the weather has been favorable, it is assumed that queens laying eggs have mated normally and have a full complement of sperm.

Worker bees clean and polish the cells in which eggs are to be deposited. A queen about to lay puts her head and forelegs into a cell and, after having determined the cell size (worker or drone), moves ahead a short distance and inserts her abdomen into the cell so that it reaches the bottom. She then deposits the egg. There is a small amount of adhesive on the end of the egg, which sticks it firmly in place and causes it to remain in a position perpendicular to the midrib of the comb. It remains in this position until it hatches.

Queen breeders assume that queens who deposit eggs which are not centered in the bottom of the cell, or are attached to the sides of the cell, are not normal. Such queens are discarded. I have seen many instances of young queens depositing more than one egg in a cell in a nucleus colony. Usually when such queens are placed in normal colonies they do not do so, and I believe that laying multiple eggs in one cell merely reflects the crowded conditions existing in the nucleus box, where there may be too few workers to clean as many cells as a vigorous young queen is capable of using.

CHAPTER 3

EQUIPMENT FOR QUEEN REARING

Queen rearing is a specialized business and requires specialized equipment. However, much of what is needed can be made by the beekeeper. One of the costly items is the mating or nucleus box. I have seen a great variation in how they are made. Since one never knows how extensively one may engage in a practice such as queen rearing, it is advisable to investigate the many ways in which nuc boxes[1] and related equipment may be made, and to settle on something that is uniform, convenient to handle and cheap to manufacture. Several types of nuc boxes are illustrated and discussed below.

Perhaps as much as anything I should emphasize the beekeepers' own convenience. Honey bees are very adaptable animals; it is we humans who tire easily and to whom time is of concern. For these reasons, some planning regarding the type of equipment to select is in order.

STANDARD EQUIPMENT

It is clear from the many experiments that have been done, and the wide variety of conditions under which bees have been kept, that honey bees will survive and prosper in a wide range of boxes and/or hives. There is no "best" type of hive as far as the bee

---

[1]Beekeepers use the word nuc to mean any small colony of bees, including that which is used for mating purposes. The word is short for nucleus.

11. This starter colony is made using an 8-frame super above a 10-frame super. Using the supers in this way allows one to remove the single piece of wood which covers a division board feeder on the right hand side of the lower hive body, to feed the colony quickly and easily. For a discussion of starter colonies, see Chapter 5. The only difference between an 8-frame and a 10-frame super is the width, as shown here.

is concerned. However, this does not prevent people from continuing to invent "new and better" beehives. As far as I am concerned, their doing so is mostly wasted effort.

What is important is protecting one's investment and making certain the equipment is interchangeable. Beekeepers must understand that at some time their equipment will be sold or exchanged. It is also likely that they will acquire additional equipment someday. About 90 percent of the beehives used in our country are of the 10-frame, Langstroth size. Nearly 10 percent are 8-frame Langstroth supers, and the rest are a mixture of odd sizes that someone, sometime, thought was a practical, standard type of beehive. I have always recommended the use of standard, 10-

12. *This small colony is being fed by placing a one-gallon widemouth jar above the hole in an innercover. It is not necessary to use an innercover, and doing so prevents one from feeding more than one jar of syrup at a time. However, in cool weather the innercover may help retain some heat in the colony.*

13. Here is a division board feeder being placed in a standard hive. The feeder is always put along one side of the colony so as not to divide the brood nest. Division board feeders may be made of any width and to hold any amount of syrup.

frame full-depth supers. Many beekeepers use half-depth supers of sundry designs, but I do not allow any in our operation. They are not interchangeable and are a nuisance. The only excuse for their existence is that they are lighter to handle when full.[2]

Our nuc boxes hold five standard full-depth frames. They work well for mating boxes. However, if I were to grow queens on a large scale I would consider using a box that would hold half-length, half-depth frames. Stocking and maintaining a large number of nuc boxes is expensive and time consuming. On a large

---

[2]The use of $6\frac{5}{8}$-inch-deep supers has become increasingly popular among some commercial beekeepers both for brood nests and honey storage. David Miksa, a Florida-Wisconsin migratory beekeeper, divides one of these supers into three parts with two pieces of masonite and uses it as a three colony mating nucleus. At the end of the queen rearing season in Florida, one of the three units is left with a queen, and the two division boards are removed. This single, queenright unit is moved north for honey production. Miksa has done this for several seasons with good success.

14. This colony is being fed three one-gallon jars of syrup simultaneously. If the weather is warm and natural nectar is not available, colonies may consume and store this much syrup in a few days. It is helpful, but not necessary, to cover the jars with a burlap bag before putting the cover and innercover in place.

scale one cannot afford an expensive 5-frame mating box such as we use. On a small scale, I feel it is too costly to construct both special boxes and special frames. I would not tolerate special 5-frame boxes, except for the fact that modifying standard 10-frame supers to hold two, three or four nucleus colonies (any of which may be done) requires building both special covers and bottom boards. The nuc box is a very special problem for the beekeeper who grows queens and believes in the use of simple, standard equipment.

METHODS OF FEEDING

When growing queens, we are especially concerned that the nurse bees who feed queen larvae have an abundance of pollen and honey so that their pharyngeal glands will be at maximum production. Feeding mating nucs is usually necessary, too, as they are seldom populated enough to gather sufficient food for their own needs. The person growing only a few queens each year may circumvent the feeding problem by producing queens only when there is a natural pollen/nectar flow in process.

Honey and pollen stored in combs is the best food for mating nucs. Some beekeepers think it is too expensive to feed honey, but this is a question of the honey quality, its availability, and time.

There are several designs for sugar syrup feeders that may be used in hives. It is never satisfactory to feed sugar syrup in bulk

15. Breakage is a problem when using glass jars. This carrying case is made from scrap plywood. We have had little breakage since we made several of these carrying boxes.

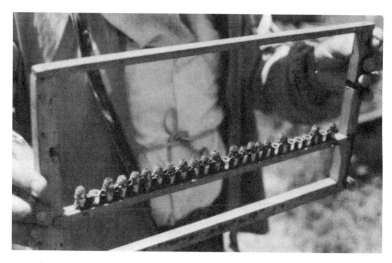

16. *This is a frame with one cell bar and queen cells, which at this stage would contain larvae about three days old. Note that the bees did not accept and add wax to all of the artificial queen cups. The cell bar fits loosely into slots made in the frame's end bars. In this way the cell bar may be removed easily for grafting or when it is moved from the starter to the finishing colony.*

17. *The bees started almost all of the cells on this cell bar. The frame is being held upside-down. The piece of wood under the frame's top bar merely fills the space and prevents the construction of comb in that place.*

18. This is a frame with three cell bars, again showing that not all grafted cells are accepted by the bees. The number of cells accepted is a function of many factors, including the amount of food available.

from an open feeder exposed in the apiary. Such a feeder stimulates robbing. Also, the distribution of food is not uniform among colonies when a bulk feeder is used.

At the present time we feed our bees using one gallon wide-mouth jars. Breakage is a problem, but we have built special carrying cases to circumvent that difficulty. Twenty to 30 one-sixteenth-inch holes are drilled in each of the bottle lids. Holes also may be punched with a nail, but drilled holes make it easier to clean the lid. One gallon jars are satisfactory for feeding our nucs, too; the bottle is covered by inverting one nuc box over another. These large jars are not satisfactory for feeding smaller nucleus colonies.

The second most popular method of feeding all types of

colonies used in queen rearing operations is to use a division board feeder. Plastic feeders that may be used in place of a standard frame are popular with many beekeepers. Wooden division board feeders may be purchased or homemade. They are made leakproof by coating with paraffin or beeswax. Small division board feeders for nuc boxes may be made by drilling multiple holes vertically in ordinary boards that have the same dimensions as the nuc box frames. Division board feeders are filled in place by using a watering can with a long spout. Large commercial queen breeding operations use garden hoses hooked up to sugar syrup tanks from which the syrup is fed by gravity or pump.

Very few commercial queen breeders use pollen substitutes during the active queen rearing season. However, they may be used effectively for a few weeks or a month before natural pollen becomes available. There is no good alternative to natural pollen for growing queens. Those who would grow queens on a commercial basis should move to areas where queen rearing is naturally productive and profitable. Persons who grow only a few queens must understand their local ecology and take advantage of it. To a lesser extent, I am disturbed that some of the pollen supplement (containing some natural pollen) is collected abroad where there is little or no control over bee diseases; it is possible that it was through the sale of pollen supplement[3] that chalkbrood (a fungus disease) was brought into this country in the late 1960s.

CELL BARS AND FRAMES

Even the growing of only a few queen cells requires some type of device to hold the developing cells so that they may be centered in the brood nest of colonies in which they are grown. Most queen breeders use a frame of standard dimensions that will hold three cell bars, which are strips of wood to which the cells are attached. Solid bottom bars such as those used on frames make good cell bars, and it is only necessary to cut them to the proper length. The precise dimensions of a cell bar are not important.

---

[3]Pollen supplement contains some natural pollen while pollen substitute does not.

A standard frame will hold three bars two inches apart. The topmost cell bar is positioned about two inches from the top of the frame top bar. Some queen producers nail shortened bottom bars above the place that holds the uppermost cell bar. Using foundation above this permanently fixed bar allows bees to build a piece of comb above the uppermost cell bar. I've never thought this was necessary, but it may facilitate the formation of a cluster around the queen cells in the event of inclement weather. It is one of those refinements one would undertake only if one had the extra time or inclination.

Small strips of wood nailed or glued to the insides of the frame's end bars hold the cell bars. The cell bars will soon become covered with a sufficient amount of wax so that they will not fall out of place. I've seen queen breeders place a strip of wood or metal along one edge of the end bars to help hold the cell bars in place, but I don't think this is necessary.

Making cells (cups)

Cell cups may be purchased from bee supply houses (the minimum order is usually several hundred) or they may be made at home. Some beekeepers cut and save the natural cups they find in colonies and use these. Honey bees are adaptable in their acceptance of queen cups and will even accept plastic cups, which are favored by a number of queen producers.

Bee supply houses also sell wooden dipping sticks for making one's own cups, though it is easy enough to fashion a dipping stick from a dowel three-eights of an inch in diameter. It is only necessary, using a fine wood file or sandpaper, to taper the tip of the dowel. The taper is made starting about half an inch from the end of the dowel. The diameter is reduced to between one-quarter and five-sixteenths of an inch at the point the tip is rounded. One may check the diameter and taper by placing the dipping stick into a natural queen cup. The dipping stick may be made of any hardwood that will not raise large fibers when it is dipped into water, as it must be when the cups are formed.

Only clean beeswax should be used to make cell cups. To

make cups the wax is melted, but care should always be taken with melted wax because it is highly flammable. The dipping stick is dipped into cold water for about 10 minutes and then into the hot wax to a depth of three-eights of an inch, removed quickly and allowed to cool. It is then dipped three or four more times, each time to a depth which is slightly less. In this manner one builds a cup with a tapered edge that will be easy for the bees to chew and modify as they extend its length. Queen breeders who grow large numbers of queens have made gadgets that will hold many dipping sticks and may be used to form 20, 30 or more cups at one time. The pan for the wax is insulated and the temperature controlled with a thermostat.

The cup is removed from the dipping stick by first dipping it

19. A ripe queen cell, on a wooden cell base, is about to be placed between the top bars of two frames where the queen will emerge. The use of a wooden cell base makes it easy to grasp the cell without damaging it, and also to fix it firmly in place in a mating nucleus. Wooden cell bases are a luxury, however, and not really necessary.

20. A piece of drain tile dug into the ground is used as a hive stand to hold this mating nucleus. One cannot afford to purchase new tile (at least on a practical basis) for this purpose, but if it is available at a low cost it makes excellent hive stands. Beekeepers have used a great variety of things like tires, bricks, and cement blocks for hive stands. It is important to keep colonies off the ground. Manipulating colonies at the height shown here is easier than working them closer to the ground.

into cold water and gently twisting it free from the end of the stick. Some commercial queen breeders, who have devices for dipping multiple cups at one time, may place them directly onto a cell bar by putting a line of hot wax onto the cell bar, forcing the newly dipped cells into the molten wax, and then removing the sticks after the bar, cups and sticks have been dipped into cold water.

A maximum of 20 cups is placed on a single cell bar. The cells are about three-quarters of an inch apart, center to center. Many queen producers place only 15 or 16 on a bar. In this way the cells are better centered in the brood nest. However, if one uses populous colonies to grow cells, the bees will usually care for as many as 20 cells per bar.

Cups made in advance should be kept free of dust by storing them in a sealed box. Wax moths are not a problem for beeswax

21. This is a shaded mating apiary in Brazil. In tropical areas it is helpful to protect the colonies from intense heat. The trees also serve as landmarks. What is not clear in this picture is that the nuc boxes are painted different colors so as to reduce drifting.

22. *This frame, on which the queen is confined for 24 hours to obtain larvae of a uniform age for grafting, is being removed and will be replaced by another dark comb in which the queen will be forced to lay during the next 24-hour period. The little mass on the second and third frames from the left is pollen supplement. The colony is being fed using a Boardman feeder; Boardman feeders work well in the South and during warm weather, but are not effective during cool weather (and often at night) because the bees cannot move to them to take the syrup.*

43

23. The two pieces of queen excluder on either side of the frame in which the queen is confined can be seen in this picture. The pieces of excluder are about two inches apart, which is wider than the normal spacing of frames in a colony. This makes it easy to remove the frame and there is less likelihood of hurting the queen. The pieces of excluder material fit into grooves in the ends of the hive body and against the bottomboard.

that has been rendered and made into cups or foundation, and one need not take special precautions to protect it. Queen cups should not be stored in the presence of paradichlorobenzene because this wax moth fumigant is volatile and may recrystallize on the surface of a cup, thereby causing bees to reject it when it is placed in a colony.

WOODEN CELL BASES

Most commercial queen producers attach their homemade or manufactured beeswax cell cups directly to a cell bar with hot wax. We twist the cell cups into a wooden cell base and attach the wooden base to the cell bar with hot wax, or by merely applying pressure and twisting it firmly onto the cell bar. This is made easier if the cell bar is coated first with a heavy layer of beeswax.

I like wooden cell bases, and for persons growing only a few or even as many as a hundred cells I advise their use. They are cheap and may be reused for many years. Some beekeepers color-code or number the wooden cell bases for record keeping. A queen cell is a delicate object. The wooden cell base allows one to grasp the finished cell without fear of damaging it or its contents. When a cell with a wooden base is placed in a mating nuc box, it is possible to press the cell into place so that it will not fall out.

Using wooden cell bases is a nuisance for commercial queen breeders who would be forced to pick up and clean large numbers of them. This is scarcely practical under their circumstances, and I understand perfectly well why they are avoided by many bee-keepers, large and small.

HIVE STANDS

I prefer to use some kind of wooden, metal or stone hive stand which raises the colony or nuc off the ground and keeps the bottomboard dry. This is especially important in northern areas where the ground tends to be damp, especially in the spring. Colonies with damp bottomboards use a larger portion of their working force to warm the hive. Raising colonies four to 10 inches off the ground also causes the entrances to be better exposed to the sunlight and prevents grass from shading or blocking them. Part of good bee management includes keeping the colony under physical conditions which will encourage its growth and development.

A second consideration for hive stands is one's own convenience. Normal colonies usually are kept on stands fairly close to the ground. Working mating nucs, which are usually several feet apart and at ground level, requires one to repeatedly bend over into an uncomfortable position. A great variety of stands that will hold the nucs at a height of about three or four feet, which is a convenient working level, have been designed and some are illustrated. The problem with all such devices is their cost. Whether or not one uses special hive stands usually depends on having a source of cheap materials from which they may be made.

GRAFTING TOOLS

Bee supply houses sell metal grafting tools that work quite well in transferring young worker larvae from their cells to queen cups. Larvae that are well fed float in the royal jelly; the spoon portion of the grafting needle or tool is easily pushed under them so they may be lifted from their cells and transferred without being injured. One may make a metal grafting tool by gently hammering the spoon portion. A thin brass rod used for metal brazing is easily worked, as is a piece of heavy copper wire. The edges of the spoon must be polished and made smooth.

One of my graduate students from Poland introduced me to the technique of making a grafting tool from a toothpick. If one is grafting only a small number of larvae, this works quite well. One holds the large toothpick in his mouth until it is thoroughly wetted; then it is easy to mold a spoon on the wetted end with one's teeth.

CAGES TO CONFINE LAYING QUEENS

To graft a few larvae, it is necessary to search through a hive until some at the right age are found. Queens usually lay eggs in

24-25. A pine board is used here as a division board to make a five-frame nuc in a standard super. Nails driven into the end of the board serve as lugs. The empty space in the hive should be filled with burlap, paper, or some other material.

adjacent cells, so adjacent brood is usually of a similar age. It is therefore not difficult to find 10 to 100 larvae to graft.

If many larvae are to be grafted into cells, it is important to confine the queen to a single comb for 12 to 24 hours. Most queen breeders confine the queen for a full 24 hours. After this time the comb with eggs is removed from the cage in which the queen has been confined and placed in an incubator colony. Sometimes it is placed just outside the cage in which the queen is laying and the same colony incubates the eggs. Three days later the larvae, which will be from an hour to 24 hours old, will be ready to graft. It is important that the colony used for egg incubation have an abundance of young bees that are well fed so that the young larvae are floating in a large quantity of royal jelly. This makes grafting much easier.

Combs placed in confining cages should be old, dark brood combs that have been placed immediately above a queen ex-

26. *This is a hive body end board with three quarter-inch-wide and quarter-inch-deep slots cut with a power saw. The slots will hold quarter-inch-thick pieces of plywood which will divide a standard super into four parts. The single hole in the lower left middle will be an entrance for one of the four units; a similar hole is drilled on each side of the hive body so that each of the four entrances faces a different direction.*

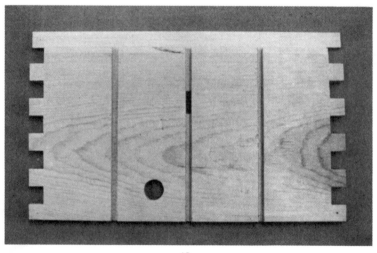

27. This is a nailed-up super divided with three pieces of plywood. Four strips of wood, each one-half inch wide, are fitted on the four undersides of the super so as to extend its depth; if this is not done, there would be no bee space under the frames.

cluder, which is immediately above an active brood nest, for several days prior to the time they are used. In this way the cells in the comb will have been cleaned and polished for egg laying and ready for the confined queen to use. I prefer to scrape and clean the combs to be used in this manner so that there is no burr or brace comb that may interfere with removing the brood comb from the cage. In a narrow cage it is very easy to kill or injure a queen.

There are several commonly used methods of making confining cages. One is to make a cage for a single comb which may be lifted from a super and placed in any hive in the same way a frame is put into position. I much prefer to cut the end boards of a standard hive body before it is nailed together, with a wide-blade bench saw, and insert two pieces of queen excluder material,

28. *The four pieces of plywood dividing the super extend above the super about one-half inch. A piece of half-inch-thick plywood is fitted in between, which serves as an innercover. In this manner, each unit may be opened and examined without disturbing the others.*

29. *This is a small nucleus colony with a piece of oilcloth for an innercover. Against the side of the nuc box is a homemade division board feeder. The feeder is made from a solid piece of wood which has been hollowed with a drill to make a space for the syrup.*

30. *This is a frame from a small nuc, well-stocked with honey and bees. This box is ready to receive a ripe queen cell. The number of bees on this and the two adjacent frames is sufficient to care for the queen.*

which have been cut to the right dimensions, into the saw kerfs. The cuts should be about one-quarter of an inch deep, the same depth as the rabbet that constitutes the frame rest. The excluder material may be cut flush with the top of the super, though I prefer to let it extend above the super about three-eights of an inch. If this is done, three pieces of plywood are cut to serve as an inner cover. One piece covers the single frame where the queen is confined and the other two pieces fit on either side. The queen excluder material may rest on the hive bottom or fit into grooves prepared for it on the bottom board.

The area where the queen is confined should be in the center of the hive, and brood should be kept on either side so as to make a more or less natural brood nest. In a standard 10-frame hive the

31. These are nuc box frames with lugs which overlap. The frames are used side by side in a nuc box; however, they may be placed end to end in a larger super when the beekeeper puts them on a colony with normal size supers. This may be done when foundation is drawn or the combs are filled with honey and/or brood.

combs are spaced just a little under one-and-a-half inches apart, center to center. In confining cages the cage sides are usually two inches apart, which leaves considerable room when combs are removed or put into the cage.

A third popular method of confining a queen to obtain eggs for grafting is to use a push-in cage made with 5-mesh (five wires per inch in each direction) hardware cloth. Hardware cloth of this size also works reasonably well as queen excluder material. In fact, if 5-mesh hardware cloth was manufactured more accurately, it could be used in place of queen excluders. Unfortunately, the hole size varies too much for it to be used in this manner. However, one can check small pieces of 5-mesh for imperfections.

Push-in cages are usually about six inches square. The sides are about three-quarters of an inch high so that they are pushed into the comb about three-eighths of an inch. This leaves a space three-eights of an inch high in which the queen may roam. Worker bees move through the holes in 5-mesh hardware cloth as easily as they do in queen excluders. Worker bees usually will chew the comb around the edge of a push-in cage and may release a queen after 36 or more hours.

Queen breeders who graft day after day may use the same queen continuously. It is only necessary to add brood to the colony to maintain the colony population, and exchange the frame in which she is laying for a fresh one, as required.

Types of nuc boxes and frames

Small colonies appear to serve better as mating nucs than large ones. By small colonies I mean those containing 1,200 to 1,500 bees and perhaps more. Why size makes a difference is not clear, but apparently the smaller the number of bees, the better the recognition and care of the introduced queen cell. If larger colonies are used for mating nucs, e.g. colonies that would fill one or two standard supers, there is a greater chance a unit will destroy an introduced queen cell and rear its own queen.

In addition to restricting the number of bees used in a mating nuc, it is also advisable to restrict the number of combs.

Two standard frames are sufficient, although as I have indicated we use nuc boxes with five frames; certainly no more than five should be used. When only a few combs are placed in the super, the empty space should be filled with a division board and burlap or paper. It is also possible to divide a standard super into two, three or four parts using one-quarter-inch pieces of plywood, which may be nailed into place.

The best way of dividing a standard super for the purpose of housing several mating nucs is to cut quarter-inch-wide and quarter-inch-deep saw kerfs into the super end boards before it is nailed together. A bottomboard is not used; rather, boards or a single piece of plywood, about 16 1/4-by-20 inches, are nailed over the bottom of the super. However, before the bottom is nailed into place the super should be made deeper by adding four half-inch wide cleats to the four bottom edges. If this is not done, the frames will rest on or close to the bottom and there will be no space under the frames for the bees to walk. Holes, usually about three-quarters of an inch in diameter, are drilled into the sides and ends of the super for entrances—there being as many entrance holes as there are units in the super (the usual number is four).

The smallest mating nuc boxes I have seen were made using three comb honey sections, each four-by-five inches. These boxes were a little smaller than I prefer and were more difficult to stock with bees. Strips of wood were nailed to the tops of the sections to make lugs that allowed the frames to hang normally in the nuc box.

As I indicated earlier, if I were to grow queens on a large scale, I would prefer using mating nuc boxes that could hold four half-length, half-depth frames. There are several reasons for doing so. One is that the beekeeper may use a standard half-depth super to draw the foundation in the combs. It is also possible to place supers of these combs on normal size hives and have them filled with honey and pollen, which may be used for nuc food. There are several sources of honey rich in pollen, such as goldenrod in the north and Brazilian pepper in south Florida, that usually do not command high prices in the market but make excellent bee food.

Feeding nucs is a time-consuming process, and using honey gives bees in the nucs a good source of food, usually at a reasonable cost.

Supers holding half-length frames may be made in several ways. They may have lugs in the center of the super which overlap or merely meet. Using frames of this design also allows one to store the combs in half-depth supers, which are much easier to stack and manipulate than nuc boxes. Queen breeders in the South find protecting their nuc box frames against the wax moth especially difficult. I know one queen breeder who uses odd size nuc box frames and places these in long, specially designed boxes which are then placed in a fumigation room for storage.

Most nuc boxes have the bottom nailed permanently to the sides of the box. The entrance is usually a hole drilled in the side. The cover is often a single piece of well-painted board that merely rests on the box, or at most has one cleat on the underside to guide the cover to the correct position.

In most areas where queens are grown, there is a greater danger of overheating the nuc box and the bees than there is of chilling it. In cool weather bees will cluster around the brood and a queen cell to keep them warm. Many queen breeders drill a hole, about an inch in diameter, in the nuc box on the end opposite the entrance. This is covered with a piece of ordinary window screening nailed into place. This makes it much easier for the bees to ventilate and cool the box during warm weather.

Chapter 4

Simple Methods for Growing a Few Queens

The honey bee is not domesticated. The methods and techniques we use today were, for the most part, designed in the late 1880s through trial and error. Today, while beekeepers agree on the general principles of colony management, one will find great variation in how individuals manage their colonies and apiaries. I think the best assessment of a management program is to make a judgment at the end of the year. At that time one can ask how many colonies he had in the spring, how many were present in the fall, and what was the total amount of honey produced.

There are many methods for growing a few queens. Some techniques work better than others. The chief goal is for the beekeeper to establish and use a technique that will work well. In this and the next chapter I will discuss a variety of methods for rearing queen bees.

Simple colony division

The easiest way to increase one's colony numbers is to divide a two-, three- or four-story colony into two or more parts, giving each a bottom and cover. As long as eggs or young larvae are present in whichever half happens to be queenless, the bees will rear a new queen. In those northern areas where the chief crop is goldenrod, which blooms in August and September, the method works quite well if the division is made in the spring.

Obviously the above method has many drawbacks. It is slow and sometimes fails. The beekeeper who checks the colony carefully, leaves the greatest number of bees and the queen on the old hive stand, and makes a queenless nuc nearby probably will have greater success. Such a beekeeper takes less risk and does not lose the parent colony. In certain years, reducing the strength of a populous colony in this manner, just before the swarming season, may be a good method of swarm control. If properly done, the strength of the parent colony is not reduced enough to adversely affect honey production in an early honey flow.

Given a choice, in most production areas I would much prefer making weak nucleus colonies, giving each of them a queen cell or a mated queen. This gives better insurance against the loss of the new colony and enables it to start the production season earlier.

CROWDING AND FORCING QUEEN CELL PRODUCTION

Congestion is the primary cause of swarming. In each beekeeping area there is a primary swarming season. For us in upstate New York it is May 15 to July 15, with a peak about June 15.[1] As one moves south the season is slightly earlier; in Florida it probably peaks in April, although we do not have precise dates for that state. In the more northern honey producing areas, including Canada, the swarming season is only slightly later than it is in Ithaca, New York.

If one wants to grow only a few queens, it is possible to select a colony with a good queen, preferably more than a year old, crowd it, and thereby encourage the production of queen cells.

---

[1] Dr. Dewey M. Caron wrote in the January, 1979 issue of *American Bee Journal* that most swarming in Maryland took place from about mid-April to mid-June, with a peak in early or mid-May. Caron had data on 402 swarms taken in 1975 and 1976. Dr. Robert E. Page, Jr. reported in the April, 1981 issue of the same magazine that 88 percent of swarming in north central California occurred in April and May with a peak about May 4. There were 121 swarms observed. As far as I am aware, data have not been collected for other areas of the United States.

Usually one may crowd a colony by not adding supers. A normal two-story colony usually will produce queen cells under these conditions; if one reduces the colony to a single super it is almost certain to do so. The best cells will be produced, when one practices this technique, within the natural swarming period and at a time when there is a nectar flow in progress.

This method of growing cells has obvious drawbacks, the chief being that one does not always know the age of the queen cells. One fact that has not yet been mentioned, but is important in the rest of this chapter and the following one, is that from the time the cocoon is spun until a day before the queen is to emerge from her cell, one must handle queen cells with great care. They must be held in their normal position, not on their sides or upside down, and must not be shaken or jarred; if cells with developing pupae are shaken only a little, the normal development of the legs, wings and especially antennae may be prevented. Often even slightly jarring an immature queen cell will kill the developing

32. These are queen cells made using the Miller method. The bottom portion of the comb was cut away exposing eggs about to hatch; two out of three eggs are destroyed so there will be a space between the queen cells. The ripe cells will be carefully cut from the comb and placed in mating nucs. Photo by R. D. Fell.

pupa. Therefore, it is important that the beekeeper know the ages of his or her queen cells.

A greater danger, of course, is that the colony may swarm, and if the beekeeper is not present, some of the bees may be lost. This may be prevented by removing the queen before the cells are capped, because a swarm will not depart before that time under normal circumstances. When the queen is removed, one should take note of the age of the oldest queen cells and then return to the colony the day before the first virgin queen is to emerge. Leaving the queen cells in a strong colony for finishing and incubation is far better than separating young cells and trusting their development to small nucleus colonies.

Some beekeepers may be concerned that producing queen cells in this manner may encourage the swarming impulse in their bees. There is a popular theory among beekeepers that one should breed only from colonies that do not swarm; in fact, it has been said, without there being any data, that one may breed bees for non-swarming tendencies. While this may be true, and some people claim to have done so, there are no data. I firmly believe that a beekeeper who is selecting for non-swarming tendencies in bees might also be the kind of careful manager who unwittingly, perhaps, uses those practices that prevent swarming. In any event, by selecting the colony(s) to be manipulated in this manner we are in no way affecting the genetic material that controls swarming in honey bees.

I am also certain that there are those who will be critical of my advocating this method of growing queens and making increases. I agree it is not very sophisticated in light of our present knowledge of beekeeping, but we must keep our goals in mind. Both this method and the simple splitting of colonies will work. Both methods take time, but otherwise they are cheap.

The Miller method of cell production[2]

Dr. C. C. Miller was a physician who gave up medical practice and devoted himself to writing and comb honey production. He claimed that he never invented anything new in beekeeping. He was a prolific and critical writer. Miller also was a careful observer whose recommendations were closely heeded.

Miller devised his own system of queen rearing. He stated clearly that it was not a method for the beekeeper who grew queens on a large scale, but was well suited to the commercial honey producer who like himself was constantly in need of a few queens.

The first step is to fix three or four pieces of foundation into a standard frame, about two inches wide at the tip and tapered to a point about two inches above the bottom bar. No wire is used to hold the foundation in place, so it is necessary to use heavy brood foundation that will not stretch or sag. The beekeeper selects the best colony (queen) and removes all but two frames of brood. The frame with the foundation is placed between these two brood frames in what would be the center of the brood nest. In about a week the comb is taken from the colony. The bees will have drawn much of the foundation and the new comb will contain eggs and larvae. Since queens lay eggs in a concentric pattern, the lower portion of the comb will contain the youngest brood. Miller makes no mention of feeding the colony during this time. He lived in a good agricultural area in Illinois. His other writings suggest there were abundant sources of pollen and nectar during much of the active season. Beekeepers who grow queens using Miller's methods in areas where food is not abundant should feed their colonies to stimulate wax secretion and comb building.

When the new comb is removed from the colony, the lower edge is trimmed so as to leave several eggs adjacent to recently hatched larvae. Two out of every three eggs should be destroyed so that the queen cells made around those remaining will be separated from each other. Those left will be eggs about to hatch. Miller

---

[2]Miller described this method in the August, 1912 issue of the *American Bee Journal.*

33. *This is a strip of comb with alternate eggs left in the single row of intact cells. This figure is from Alley's 1885 book on queen rearing. Cutting the comb in this fashion is the first step in making cells according to the Alley method.*

34. *This is a second figure from Alley's book. The strip of comb shown in the figure above has been stuck onto the underside of a comb with hot beeswax. Separate and distinct cells may be produced in this manner. They are cut from the frame when they are ripe, as with normal cells, and placed in mating nucs.*

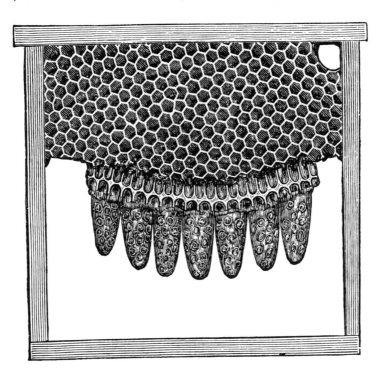

called this frame "queen-cell stuff." This frame is placed in the middle of the brood nest of a dequeened colony. Miller says the bees will do the rest. Apparently bees prefer to draw cells from the newer comb, because Miller states they will not draw cells (or at least not very many) from the darker comb in the queenless hive even though larvae of the proper age are present. About 10 days after the comb is introduced into the queenless hive, it may be replaced and the ripe queen cells cut away. These are placed in nucs, where the virgin queens will emerge and from which they will fly to mate.

The Alley method of cell production

"In order that we may be successful in the production of honey it is of vital importance that our queens are first-class in every respect."[3]

Alley confined his breeder queens in small colonies with five combs "4.5 by 5 inches." "If" a comb is "placed in the hive at night it will be filled with eggs by the next morning." On the fourth day the eggs will have hatched and the larvae will be of the proper age for starting cells. It is best to use a piece of new comb that may be cut into strips easily, as described below.

The next step is to prepare the cell building colony. Alley advised using the best colony in the apiary. This colony is smoked, the bees confined, and the hive drummed for about 10 minutes to cause the bees to engorge. The queen is then found and the queenless bees shaken into a swarm box—a super without combs with a screen on the top and bottom. Alley suggested it was best to shake the bees into the swarm box in the morning and give them the frame with day-old larvae in the evening. The bees should be queenless for at least 10 hours before being given the larvae for queen rearing.

The next step is to cut the comb, containing the proper age larvae, into strips by cutting through alternate rows of cells length-

[3]So wrote Henry Alley in *The Beekeepers Handy Book: or Twenty-two Years' Experience in Queen Rearing*, published by the author in Wenham, Massachusetts in 1885 (269 pages).

wise and destroying every other larva. A strip as cut by Alley, with every other larvae destroyed, is illustrated in Figure 33, taken from his 1885 book. This strip is attached, with the cells trimmed and facing downward, to a convex cut comb shown in Figure 34, also from Alley's text.

The cell building colony, despite the fact that the bees are engorged, should be fed. A good cell building colony may start up to 25 cells, but Alley advocated destroying some so that the colony would finish only 12; he considered this the maximum number a colony should be forced to care for.

The comb with the prepared larvae is placed into an empty super with other combs. This is placed over the swarm box, or the bees in the swarm box are dumped in front of the super so that they will enter the cell building colony. The following morning the bees will have free flight but will concentrate on cell building. They should be given no brood. After the cells are capped a single colony may be used to care for several frames with cells. When the cells are near emergence they are cut apart and placed into mating nucs.

Alley's methods are good and will work. Swarm boxes are still used but in a manner slightly different from that described by Alley. His mating nucs also were different, but it must be remembered that Alley was the first to develop a practical method for queen rearing. We cannot expect his original methods to be perfect.

CHAPTER 5

COMMERCIAL QUEEN PRODUCTION

Modern methods of rearing queens in large numbers were developed near the end of the last century. Henry Alley of Massachusetts, whose methods were described in Chapter 4, was the first to develop a practical method of rearing queens. However, it was G. M. Doolittle of Borodino, a small village southwest of Syracuse, New York , who is considered the father of our modern queen rearing industry. Doolittle's book, *Scientific Queen Rearing*, was first published in 1888 and was reprinted several times.

While Doolittle did not invent artificial queen cups, he was the first to understand that they could be made and accepted by bees. His grafting tool was whittled from a toothpick. Doolittle also was the first man to demonstrate that queens could be reared in queenright colonies. Doolittle's apiaries were in the lime-rich soil of central New York where there are an abundance of nectar- and pollen-producing plants. In his writings and lectures, Doolittle emphasized that good queens may be reared only by simulating the swarming and/or supersedure situation, and that an abundance of food must be available at all times for the bees in the cell building colonies.

GRAFTING

Larvae of the proper age for grafting are obtained by confining a breeder queen in a cage, as described in Chapter 3. When it is time to graft, the comb with larvae—a few nearly 24 hours old—is lifted from the colony in which it has been stored and the bees removed by brushing. Queen breeders do not approve of shaking bees from combs. Larvae are delicate animals. All the larvae in the comb will be in the same instar or larval growth stage; thus they will be the same size despite their slight difference in age.

Grafting the larvae from the worker comb to the queen cells should be done rapidly and under good conditions. We prefer to graft when the temperature is above 75°F and the relative humidity above 50 percent. Day-old larvae are very difficult to see, so one needs good light that will penetrate to the bottom of the worker cells. Sunlight is good, but one must not allow the larvae or the royal jelly to dry, as they may if exposed to too much sun. Some types of lamps generate too much heat and must be avoided.

It is best that only a few steps separate the colonies from which the young larvae are removed, the grafting room, and the

35. Grafting. The grafter holds a cell bar in his left hand and reaches for larvae with his right. The frame contains mostly capped brood, so obviously only a few larvae are available for grafting. A florescent light between the grafter and the frame shines into the cells, making it easier to see the small larvae.

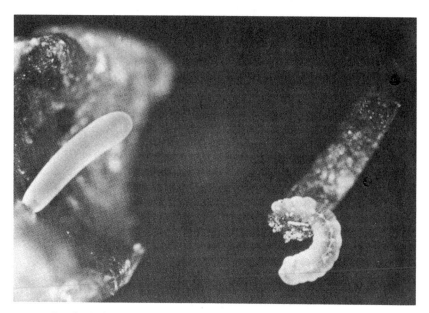

36. On the left an egg is about to hatch. On the right is a first instar larva of grafting age resting on a grafting needle. While larvae of this age are delicate, they may be transferred from their cells to queen cups with ease.

colonies about to receive the cells. Even with this precaution, we cover the rows of larvae yet to be grafted with a damp towel and also cover bars of grafted cells in a similar manner until there are sufficient numbers of bars to put into a starter colony. Some queen breeders have designed special carrying boxes for frames of larvae and grafted cells. These may be valuable, but I prefer having everything in close proximity so that a box is unnecessary.

Many queen breeders "prime" queen cells with a small drop of a mixture of half royal jelly and half warm water before the larva is placed in it. Queen cells with larvae about three days old have a maximum of royal jelly; a few such cells will supply a large quantity of jelly. When the water is added the mixture should be stirred carefully to avoid any lumps. Whether or not it is really necessary to prime cells is a debatable question. I know of many breeders who do not prime cells. They make certain the colonies caring for the young larvae are fed so the larvae are surrounded by a surplus of royal jelly. Of course, the larvae should be well fed anyway; it is very difficult to graft (pick up) a larva that is not surrounded by a great deal of jelly. The grafting tool should have a large spoon so

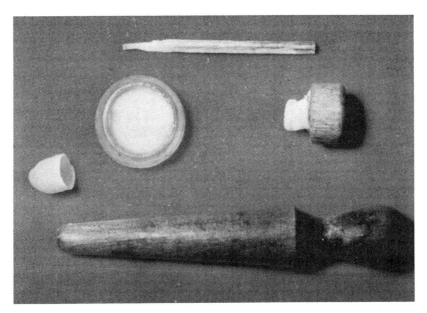

37. These are tools and equipment for grafting. Along the top is a homemade, wooden grafting needle while the tool at the bottom is for dipping the queen cups. The three items in the middle, starting from the left, are an artificial queen cup, a dish of royal jelly (used by some beekeepers to prime queen cups before the graft is made) and a queen cup in a wooden cell base.

that as much jelly as possible is transferred with the larvae. Last, the newly grafted cells should be rushed to well-fed starter colonies where the bees can feed them lavishly. If the cells are primed, it is important that the larvae not be immersed in the royal jelly but floated off the grafting spoon in the center of the drop. Larvae breath through openings (spiracles) that line the two edges of their bodies; they can drown if deprived of contact with the air.

Beekeepers often have asked if it would be worthwhile to graft eggs rather than larvae. It has been suggested that larvae hatching from eggs already in queen cups might be better fed than grafted larvae. The subject was researched by Mr. Stephen Taber III, who devised a special forceps for transferring eggs.[1] About 75 percent of the transferred eggs hatched and queens were produced from some of them. Taber stated there are no data to support the theory that queens produced by moving eggs are in any way superior to those produced by normal grafting methods.

_____

[1]Taber described these forceps in the April, 1961 issue of the *Journal of Economic Entomology* under the title "Forceps designed for transferring honey bee eggs."

DOUBLE GRAFTING

Double grafting is a little-used system in which the larvae grafted into cells are discarded after 24 hours and replaced with new larvae of grafting age. The thought behind this method is that second larvae will be better fed and a better queen will develop. It is interesting that despite the fact that double grafting has been known for a long time, no researcher, so far as I am aware, has studied the method under controlled conditions to determine its value. I suggest that double grafting is not worth the extra time involved; good results can be obtained without resorting to this method.

STARTER COLONIES

A starter colony is one in which newly grafted cells are placed. Some queen breeders use the same colony both to start and finish cells, but this is probably not advisable unless one is rearing only a small number of queen cells. Starter colonies are prepared in slightly different ways by different breeders. Some methods are as follows, the most common one being cited first.

A strong two-, two-and-a-half- or three-story colony should be selected as a starter colony. For three days, it is fed syrup made

38. This is grafting in Brazil. This grafter works without a special lamp and sits in the doorway of a small van. The cell bar is held with the left hand. On the floor of the van are other cells to be grafted.

by mixing sugar and water, 50 percent each, by weight or volume. At the end of this time the colony is dequeened, reduced to one or one-and-a-half supers and given two or three frames of young larvae. Some breeders give the colonies one of the frames of hatching eggs that will be used for grafting the following day. Feeding of the 50 percent syrup continues; the colony also should have one or two frames of pollen. The remaining brood is removed and given to another colony. The following day the frames of larvae are removed about an hour before the colony is to receive the cells. In fact, one frame may be used to obtain larvae for grafting and some of the same larvae returned to the colony, but this time in queen cups. A single strong starter colony may be used to start 100 to 120 cells for up to four successive days. The quantity of royal jelly and the sculpting of cells should be carefully examined on the third day because sometimes the care on the third day is less, in which case the colony should not be used again.

A variation on this theme is to confine the bees to a swarm box. A five-frame swarm box would contain five to seven pounds of bees, which are left broodless and queenless for four to eight hours before receiving the newly grafted cells. The swarm box has a regular cover, but the bottom is screened for ventilation. Frames in the box should contain pollen; feeding continues as the cells are being started.

Queenright starter colonies may be used, but more frequently such colonies are both starters and finishing colonies. They may be two- and three-story colonies. The queen is confined with an excluder to the lowest super and the cells are started and reared in the topmost super.

The chief problem with queenright starter and starter-finishing colonies is preventing them from swarming. This is done by using young queens, preferably those one to four months old, clipping their wings, and searching and removing queen cells that may be started by bees in a queenright portion. The ideal situation is to keep capped and hatching brood in the lowest chamber with the queen and hatching eggs, and growing larvae above with the developing queen cells. Still, one must not allow too many developing larvae near the queen cells, or nurse bees will be diverted from the cells. Combs with larvae should be removed when newly grafted cups are added every three or four days. Hatching eggs and young larvae also should be removed from the lower to the top-most super every three to four days. At the same time, capped brood—especially that near emergence—should be put into the super with the queen. The importance of feeding and pollen has already been discussed.

FINISHING COLONIES

Finishing colonies are more frequently queenright than queenless. Swarming or attempted swarming is a problem for queenright hives; in the case of queenless colonies, one must continually supply frames of emerging brood. Finishing colonies usually are given two or three[2] bars of cells every three or four days, fewer than are given to starter colonies.

In finishing colonies, which may be two or three stories, the queen is confined below and developing worker larvae are kept above to attract the young bees to the vicinity of the cells. As in the case of queenright starter colonies, the queens must be young and have their wings clipped and the colonies must be continually checked for queen cells in the queenright portion.

CARE OF CAPPED CELLS

While the developing queen is in the larval stage, she floats in the surplus of royal jelly in the open cell. While it is important that this cell not be jarred or shaken severely, the larvae are no more delicate than are worker and drone brood in the larval stage. However, soon after a queen cell is capped, the larva stretches lengthwise in the cell and spins her cocoon. During this stage and until about 24 hours before the virgin queen emerges from the

[2]Some would argue that three cell bars are too many, just as they would say that a starter colony should be given fewer than 100 to 120 cells. The fewer the number both receive, the greater attention the cells will be given. Giving a colony too many cells to care for will result in smaller cells and poorer queens. These are instances in which the beekeeper, growing only a few queens for his own use, has a clear advantage.

cell, great care must be taken not to jar or disturb the developing queen. Queen cells placed on their sides during the pupal stage will die, or the virgin queen may emerge with deformed legs or wings.

Ripe queen cells—the term given to cells that will emerge within 24 hours—may be handled without fear of damage to the resulting queen. I have observed queen breeders cutting the ripe cells from the cell bars with a knife and placing them on their sides, often carrying them in a car for miles to a mating apiary. Of course, a careful queen breeder does not misuse his prerogatives. Most have carrying boxes that are adequately lined to protect the ripe cells when they are transported.

INCUBATORS AND INCUBATOR COLONIES

Colonies into which frames of eggs, 0 to 24 hours of age, are placed are sometimes called incubator colonies. However, in this section I am referring to colonies or boxes in which capped queen cells are placed and held until shortly before the virgin queens emerge. There is a prejudice on the part of some queen breeders against using incubator boxes. I have never had such an objection because I have seen several successful queen breeders use and favor them. Certainly they are cheaper than large colonies of bees, although a cautious queen breeder uses both with care.

The incubator colony is usually a cell finisher that contains too few bees to finish another batch of queen cells. While such a colony may be able to finish only 40 to 60 cells successfully, it may be capable of caring for three to five times that many capped queen cells.

The incubator boxes may be made to hold many hundred or even over a thousand queen cells. As far as I know, all those currently in use are homemade, usually extensively modified from refrigerators or other insulated boxes. It is a simple task to build in several modifications. One is the division of the box into several compartments, so that should a virgin queen emerge earlier than expected, the damage she can do attacking other cells is limited.

The greatest danger in using an incubator to hold queen cells is that the thermostat may malfunction, causing the temperature to go too high. One beekeeper told me he uses two thermostats wired in series as a safeguard. Queen cells, like worker and drone brood, can withstand a slightly lower temperature for short periods of time, but overheating will rapidly kill the developing queens.

CHAPTER 6

MATING AND MATING NUCS

Mating nucs are the small colonies into which ripe queen cells are placed; the queens which emerge are kept in mating nucs until they are mated and have begun laying. The construction of mating nuc boxes is discussed in Chapter 3.

The mating nuc is an important part of every queen rearing operation. Because nucs are usually small they are difficult to manage, and the smaller they are the more problems there are. However, making large mating nucs is expensive in terms of equipment, food and bees. Some queen breeders, especially in the South, do not use baby nucs during the warmer parts of the year but find it necessary to revert to larger nucs, usually with standard frames. Proper ventilation is often a serious problem and the chief reason for using larger nucs.

An ideal mating nuc contains bees of all ages so that all of the chores that need attention, including cooling and ventilation, feeding the young, etc. are properly attended to. Under ideal conditions a young queen will help sustain the nucleus population by laying some eggs; however, even when this occurs it is usually necessary to add bees (or brood) to the nucs as the season progresses.

STOCKING NUC BOXES

There is no best or perfect way to prepare nuc boxes to receive queen cells. If there was, almost everyone would use it, as is true with so many other management practices. I have outlined some of the more common methods below.

In our own operation, where we rear fewer than 100 queens in an average year, we use four-frame nuc boxes with full-depth frames. We stock each of these with one frame of honey (and an additional frame with some pollen if there appears to be little pollen in the comb with the honey), one frame about one-quarter full of brood[1], one empty comb, and the bees shaken from two or three other frames so that the brood will be well cared for and fed properly. This is a fairly expensive unit with regard to brood and bees, but it is also one in which we may hold a queen for a month or more after she is mated. We like to have a small reservoir of queens at all times.

Commercial queen breeders have a variety of methods for stocking small mating nucs. One method is to shake several pounds of bees into a package and feed them heavily so that they may be ladled into the nuc boxes. Some method is needed to measure the number of bees, and a large soup ladle works very well for this purpose. Just dumping a quantity of bees into a box is not accurate or satisfactory. About a third of a pound of bees (1,200-1,500 bees) is a good number. It is also ideal to have a comb of honey; if this is not available, then the nuc may be given sugar syrup. A ripe cell is added immediately after the workers are put into the box; the bees in the package should have been queenless for several hours. The nuc box entrance is closed, usually with a screen. The filled nucs are stored in a cool, dark place for three or four days, by which time the virgin queen has emerged and aged sufficiently so that she has started to produce the chemical substances by which she is recognized. At this time the nucs may be taken to the field, preferably in the evening or very early morning,

---

[1]A standard Hoffman frame has a total of 6,800 cells. A frame one-quarter full on each side would contain over 1,500 potential workers, far more than is needed to sustain the colony until the queen starts to lay.

so as to discourage their flying immediately and drifting (see below for a further discussion of drifting).

A more expensive and little-used method for stocking nucs is to place small half-length, half-depth frames in colonies with laying queens so that eggs will be deposited in them. One such comb with brood is placed in each mating nuc. The frame will have some adhering bees, but not enough to care for the brood. Additional bees from the same colony may be shaken into the nuc, or well-fed bees from a package, as described in the paragraph above, may be used. Honey or sugar syrup is added.

It is also possible to stack supers of small or half-depth frames on a colony, above an excluder, and some bees will move upward and start cleaning and settling on the combs. These may be removed with the adhering bees and placed in nuc boxes. This method requires more labor and is not too popular.

FEEDING NUCS

While it is ideal to have nucs with bees of all ages and in sufficient numbers so that foraging for pollen and nectar will take place, this does not always occur. If one provides an abundance of honey and sugar syrup, those bees of foraging age will be able to spend their time gathering pollen. Developing queens must consume a small amount of royal jelly as their ovaries grow and they begin making eggs. While worker bees can produce some royal jelly using body reserves, they must eat pollen to produce it in quantity. It is estimated that a cell full of pollen is required for each worker bee a colony rears; thus, if a nuc is expected to rear brood it will need to collect pollen. I already have commented that I am not too happy about the use of pollen supplements, but I am aware that they are used by some queen breeders with apparent success.

Using feeder jars or pails for baby nucleus colonies is difficult because of their small size. When combs of honey are not available, the next best feeder is the homemade division board feeder. It must be filled frequently. I much prefer having the bees empty the feeder between feedings, ripen the sugar syrup, and store it in combs than I do leaving the feeder half full. Mold, yeasts and fungi may grow in sugar syrup, giving it a foul odor and making it unpalatable. This may occur after as few as four or five days. A strong nucleus colony will be able to clean the feeder periodically and thereby deter the growth of these undesirable microbes. As discussed above, the ideal situation is to have combs of honey and pollen filled in advance by normal colonies.

DRIFTING OF WORKERS

The worker bees taken from a mature, normal colony will vary in age and include some foragers. If these bees have an opportunity to fly from the nucleus colony immediately, they may return to their parent colony or become lost and drift to another colony, causing an imbalance in populations. Some drifting is avoided by feeding the bees heavily and confining them to the nuc box for several days. It is also good practice to make up nucs from apiaries several miles away from where they will be used.

The standard techniques used to prevent drifting may be employed to good advantage. These include marking nuc boxes different colors, painting designs around entrances, facing entrances in different directions and making sure there are landmarks such as trees and bushes in the apiary. Queen breeders with whom I have discussed the matter have agreed that it is worthwhile to take precautions to discourage drifting.

I once visited a mating apiary where there were over a thousand nucs, each four feet from the next. The queen breeder was using markers, especially trees, to reduce drifting, which he said was a serious problem. When I asked him why he didn't put the nucs further apart, he said that if he doubled the space between the nucs, the apiary would occupy four times as much space. This would increase the walking and working time a great deal. He said he had tried different methods and his personal arrangement was satisfactory, although there was some drifting.

DRIFTING OF VIRGIN QUEENS

One commercial queen breeder told me that he was satisfied if 60 percent of the cells he put into nucs resulted in salable queens. This may appear to be a low survival rate, but there are many problems when dealing with such a small unit as a mating nuc. Drifting and the resulting loss of virgins is one of the most serious problems.

In large commercial mating apiaries one often can see small clusters of bees, sometimes 100 to 2,000 bees, hanging from a bush or tree. If these groups are examined it will be found that they often contain one or several virgins or very young mated queens. One queen breeder told me he once found eight queens in such a mass. These bees and queen(s) are rarely absconding swarms, though this, too, may occur on occasion. These clusters, queens included, are made up of lost and drifting bees. We have long known that natural swarms are attractive to and may be joined by foraging bees.

Clusters of bees found in trees and bushes in mating apiaries should be removed and destroyed unless one knows they are of very recent origin. Queens of mating age are in a physiologically delicate stage. If they are unduly exposed to inclement weather, or their mating is delayed, they may not become normal queens. If clusters are not removed, the bees in them may attract other workers and mated queens that may otherwise find their own hives.

There are no special precautions, other than those already mentioned to reduce drifting by workers, that one may use to reduce virgin queen drifting. Honey bees can distinguish blue, blue-green, yellow and white. They also can differentiate certain designs. All of these factors must be used to the best advantage.

CHAPTER 7

CARE, STORAGE, AND SHIPPING OF QUEENS

There is a strong belief among certain beekeepers that home-grown queens are best and that shipping mated queens through the mail is harmful to them. The first part of this statement may be correct, but there are no data to back up the second thought. In fact, the long history we have of successfully shipping queens through the mail suggests it does them no harm whatever. A queen stops laying eggs when her colony swarms or is forced to abscond; neither situation appears to cause the queen difficulty in further egg production.

If home-grown queens are better, it is probably because a person rearing only a few queens gives them considerable attention, especially with regard to feeding them while they are developing. It is certainly reasonable to assume that any animal well cared for will be a better producer.

BANK COLONIES

Queen breeders who follow regular schedules often have more queens than they can use or ship immediately. They may need to remove mated queens from mating nucs to make room for ripe cells. Mated queens may be removed from mating nucs, caged in regular queen cages without worker bees or queen cage candy, and placed with other similarly caged queens in a queen bank colony (a colony in which many queens may be held until they are needed). In this manner queens may be kept for many weeks and remain in good physical condition, ready to be shipped or used at any time.

The key to having a successful queen bank colony is giving the colony a frame of emerging worker bees about every five days. Some queen breeders give new brood less frequently, but I think every five days is a good schedule.

Our research has shown that older worker bees are likely to be antagonistic toward queens who are not their own. They will clamp onto such queens and hold them tightly (this is called balling[1]). Young worker bees show no antagonism and will feed all queens that solicit food—thus the importance of keeping a large supply of young bees by adding brood frequently.

Most queen bank colonies are not queenright in the normal sense; that is, they do not have free-roaming queens. However, a queen breeder told me that he thought banked queens fared better in queenright colonies. No one has studied the matter. It is possible that the presence of brood in all stages is helpful, or that constantly emerging brood may encourage the feeding of multiple queens under these circumstances.

Many people have attempted to hold queens from fall until spring in varying types of queen bank arrangements. An effort which has experienced moderate success used a system which added young bees in sufficient quantity so that the banked queens were well fed.

Finding queens

There is only one reason for including a paragraph on the subject of queen finding, and that is to state that, if possible, one does not smoke a colony or nuc in which he or she wishes to find the queen. Smoking calms worker bees and causes many of them to put their heads into cells of honey and engorge. However, this

---

[1]Balled queens are rarely killed; when they are, I believe it is in error and not done purposefully. Despite statements to the contrary, I am convinced that balling is a process whereby bees hold a queen who is not their own in the event that she might be needed at a later time.

varies with the race of bees. Some react by running while others form large clusters. Queens, like workers, react in different ways to smoke, but usually they run aimlessly over the combs and away from the brood area. This makes queen finding difficult. If the weather is favorable and one moves slowly, one often can open a nuc box and remove the combs without smoke. When this is done the queen usually will be found walking over those combs with brood; most often she is in the center of the brood nest where she can be found with ease.

CLIPPING QUEENS' WINGS

Clipping a queen's wings will not prevent swarming, but it will delay it. Bees, of course, are not aware that their queen cannot fly when her wings are clipped and they will attempt to swarm anyway. Usually what happens is that the swarm will exit, but will return when not joined by the queen. The queen remains at the colony entrance, attempting to take flight until eventually she moves back into the hive (or is lost). The swarm may attempt to leave several times, and will finally succeed when the first virgin queen emerges and is ready to take flight. During all this time the queen lays fewer eggs, and a large percentage of the workers remain engorged and will not forage, all of which adversely affects colony productivity. Occasionally, a queen with clipped wings will crawl away from the hive for a short distance and be lost. The sudden cessation of egg laying may or may not prevent swarming.

Still, it may be worthwhile for some beekeepers to clip their queens' wings since it does delay swarming. A few beekeepers clip alternate wings in alternate years so that they can determine the queen's age. Still others clip wings in various ways (diagonally, or one shorter than the other) so as to mark queens of different backgrounds.

Beekeepers agree that clipping a queen's wings does her no physical harm. Grasping a queen to hold her still so that her wings can be clipped also may be done without harming her. However, I suggest that the clipping be done rapidly to ensure that no accident occurs. I have never had a queen I was holding attempt to sting me, and I have heard only rare reports of their doing so.

To clip a queen's wings I grasp her thorax from the head end, thus exposing the wings. In my opinion it is not advisable to grasp or squeeze the abdomen or head. The chief problem I have in grasping queens is keeping my fingers free of propolis, a problem that is worse today than it was many years ago because of the extensive use of Caucasian bees in this country. One may practice wing-clipping using workers, but they are notoriously quick to curl their abdomens and sting one's fingers. It is better to practice using drones.

I have seen gadgets one may place over a queen in order to cage and hold her firmly in place while her wings are clipped and/or she is marked. I have never used these queen holders, and while I have no objection to them I do not think they are necessary. Once one has learned to catch and grasp a queen properly, I think one will agree it is the best method.

One needs a very good pair of scissors to clip wings. Those with small blades work best. We find it is helpful to set aside a pair for this special purpose. I have seen beekeepers who cut queens' wings with a penknife pushed against the thumb, but I have never approved of this practice. It is too easy to tear a wing at the point where the wing is attached.

Marking queens

We often mark the queens we are using experimentally, and some beekeepers also may find it worthwhile to do so. I know of no commercial beekeepers who routinely mark their queens, although a few may mark breeder queens.

The best way to mark a queen is to clip her wings in a distinctive way. Even the very best paints and markers may wear or fall off.

Markers on bees usually are placed on the top of the thorax between the wings. Occasionally, marks are placed on the top of the abdomen.

Probably the best queen (and worker) marker is made by mixing a pigment with shellac. Shellac is an insect secretion dissolved in methyl (=denatured or wood) alcohol. New shellac is too thin to be a good glue, but if one allows a can of shellac to sit on a shelf for a year or more, the solid contents will settle and the thin, top portion may be poured off, leaving a viscous shellac that will mix well with most dry pigments. If a small drop of pigmented shellac is put onto the top of a queen's thorax, it usually will remain there for her lifetime.

Small plastic numbered and colored discs are available from some bee supply houses. There are five colors, each with numbers 0 to 99. Glue is provided with the markers, but partially dried shellac, described above, will serve the same purpose. The discs are concave and designed to fit a queen's thorax. These same discs have been used to mark worker bees and other insects; however, they are made specifically to fit on a queen.

Most fingernail polishes are good markers for bees. Like shellac, they should be allowed to settle for six months or a year and the topmost liquid poured off and discarded. Most fingernail polishes dry rapidly.

Queen Cages

Cages for shipping queens through the mail are made by bee supply houses in a variety of sizes and shapes. Most are made of wood, but some are made of plastic. The most popular cage is made of basswood, which takes nails easily, and has two or three overlapping holes about an inch in diameter and three-eights of an inch deep. The open face is covered with wire screen (window screen). Two of the three holes, or compartments, are used to house the queen and her attendants, while the third is filled with queen candy to provide food while the bees are in transit. Several queen cages may be nailed together with strips of wood when more than one queen is being shipped to a single destination.

Queen Cage Candy

Queen cage candy is made by kneading starch-free confectioner's sugar into warm honey or invert sugar. In the trade this is referred to as "Good" candy, after I. R. Good who introduced

39. This is a queen and workers in a standard queen mailing cage. The queen's abdomen is seen in the lower right. The hole on the left is filled with queen candy.

the method from Europe into the United States late last century. Most beekeepers frown upon using honey in queen cage candy because this is one method of spreading American foulbrood, should the honey contain any of the bacterial spores.

The reason for using invert sugar or honey is that both are sweeter than a sucrose sugar syrup and the candy is therefore more attractive to the bees.[2] The use of starch-free confectioner's sugar is recommended because the starch might not be as digestible as the sugar; however, there are no data to support this contention. The honey or invert sugar usually is heated to 130-150°F before the kneading process starts in order to decrease its viscosity and speed up the mixing of dry sugar and liquid.

Making queen cage candy is a slow process. The final product must be firm and should not run in the queen cage. Newly made queen cage candy is allowed to stand for a day after it is

---

[2]Invert sugar and honey both contain fructose which does not form hard crystals like glucose and sucrose do; therefore, candy made with fructose will remain soft and easier for bees to eat. The use of honey in making queen cage candy is illegal in many states because this is one way American foulbrood may be spread.

molded into a ball. If the ball flows or slumps overnight, then additional confectioner's sugar is needed. This can be a problem especially in warm, humid weather. Queen cage candy can be stored in an airtight container for long periods of time.

When wooden queen cages are used, the hole into which the sugar candy is introduced should be dipped into or painted with hot paraffin or beeswax. If this is not done, the wood may dry out the candy, making it useless to the bees in transit. Some shippers also will cover the exposed candy under the screen with waxed paper. Bees that eat queen cage candy must liquefy it by regurgitating honey or water from their honey sacs. However, if the candy is too dry it is extremely difficult for them to do so. On the other hand, candy that is too moist may run in the queen cage, wetting the bees and perhaps killing them. There is no source of commercially made queen cage candy and beekeepers must therefore make their own. Making proper queen cage candy is one of the many aspects of successful queen rearing that deserves close attention.

MAILING QUEENS

Wooden and plastic cages have bee used to ship queens for many years. They are equally satisfactory. Six worker bees in a cage with a queen appears to be the preferred number. Queen cage candy usually occupies about one-third of the cage's space. While our post office appears reluctant to ship package bees, they seem glad enough to accept queens for delivery anywhere.

When multiple queens are shipped, the cages are stacked so that the screen faces are not in contact with each other. The wire screen on the exposed face of the outside cage is covered with a piece of cardboard.

Queen cage candy contains little water. While it is not necessary to give water to queens in transit, it is a good idea to place a drop of water on the screen of a queen cage as soon as it is received. Queens are usually caged and shipped on the same day so there is no need to give them water before they are mailed.

It is best to put queens into colonies as soon as they are

received. However, beekeepers often keep reserve queens on hand for a week or more. Caged queens should be kept in a dark place where it is not too warm and the temperature does not fluctuate. They should be given water probably every other day—more often if they take the water rapidly when it is offered.

Battery boxes

For the past decade it has been popular to ship queens in cardboard containers called battery boxes, but also known as package banks or queen nurseries. Beekeepers who have used battery boxes say that the queens arrive in better condition than when shipped in individual cages with attendants. Battery boxes may contain 100 or more queens individually caged without attendants. The queen cages are surrounded by 1,000 to 2,000 free-roaming workers.

Battery boxes are built to hold nearly 200 queens, but few shippers place this many queens in a box because the post office will not insure packages of this size for their full value. The boxes are sent by express mail if the recipient is more than 300 miles away, but by priority mail if the queens are being sent less than 300 miles. Express mail has been a great boon to beekeepers, and they agree that queens arrive in much better condition when shipped by this rapid method.

The worker bees for battery boxes are taken from the brood nests of colonies that have a quantity of emerging bees. This ensures a large number of young bees. If drones are present they do no harm. Since battery boxes are a recent innovation, changes are still being made in their design, especially in the use of plastic queen cages which also are becoming increasingly popular. The most popular battery box has two pieces of 8-mesh hardware cloth on each end for ventilation. The hardware cloth pieces are about one-quarter of an inch apart. In this way there is little chance that anyone handling the box will be stung through the screen. It is not necessary to water the bees en route. A small glob of soft sugar candy is pressed against the cardboard for food while the bees are being shipped. One may tape the food in place, but this is usually not necessary.

INSPECTION CERTIFICATES

Most states, including all states in which queens are produced, have a state apiary inspector or chief entomologist to oversee apiary inspection. The chief concern for apiary inspection departments has been American foulbrood, a contagious bacterial disease which affects only bee brood. In recent years, apiary inspectors have been alerted to check for new bee diseases and parasites which may have been introduced from abroad, and to deter the interstate shipments of new diseases caused by mites.

Most states require that an apiary inspection certificate accompany each shipment even if only one queen is being shipped. This is often printed on the piece of cardboard covering the screened face of the queen cage. While an inspection certificate does not guarantee the bees within the cage are disease free, it is a good indication that this is the case. The matter need not be of concern to beekeepers receiving queens, as I have never seen a shipment of queens without a certificate; queen breeders have been very careful to make certain the certificates comply with the law. A national association of state apiary inspectors, the Apiary Inspectors of America, meets once a year to review new and needed legislation. This has been a very effective organization working on behalf of beekeepers.

Those who intend to sell queens should consult their state department of agriculture concerning the rules and regulations pertaining to bees both within their state and in those states to which they expect to send queens.

CHAPTER 8

REQUEENING COLONIES

Commercial beekeepers may be classified as those who operate extensively and those who do so intensively. The latter have fewer colonies and make certain each one is a producing unit; these beekeepers usually requeen their colonies every year. The first group own far more bees and usually depend upon the colonies undergoing natural supersedure; not every colony is expected to produce a maximum crop. Both types of operation have been successful. In areas where honey flows are more erratic, one is inclined to operate more extensively.

However, given a choice, I prefer to requeen annually and attempt to make each hive produce as much honey as possible. Requeening large colonies, as discussed below, is not an easy task.

WHEN TO REQUEEN

In the northern states, those who routinely requeen usually do so on or about August 1. The chief concern is to have a young queen that will lay late into the fall and initiate brood rearing early in the spring. The first of August is selected over late August or September because it is usually the time between the early and late honey flows. Also, if requeening fails at this time, it is possible to make another attempt so that the colony still will have time to prepare for winter and not be lost. This recommendation, of course, pertains to the northern states and Canada. Beekeepers farther south probably would requeen later in the year, keeping in mind that it is the spring buildup which is usually most important for honey production.

Many commercial beekeepers keep a small number of single-story nucleus colonies in their apiaries at all times, in the event that a queen fails during the active season. These colonies usually are started in the early spring, using a single frame with some brood and sufficient bees to cover, together with a new queen. If such colonies are allowed to grow unchecked, and there is a good fall honey flow, they often will become sufficiently strong by September in order to winter without further feeding or assistance. If these reserve colonies are too weak to winter by themselves, it is often possible to combine two such units about September 1 and obtain a colony which may winter.

Some beekeepers make up as many as ten percent reserve nucleus colonies in each apiary to be ready for emergencies. It is often possible to save a mature colony that has accidentally lost its queen, or is otherwise deteriorating, by requeening with one of these nucleus colonies.

Estimating a queen's worth is best done by checking her brood pattern. This is not infallible, and one must be familiar with all of those things that can go wrong and affect a brood pattern. This is discussed in greater detail in Chapter 9. However, it is important to remember that any time a queen begins to fail she should be replaced, even though she might be a relatively young queen.

METHODS OF QUEEN INTRODUCTION

There is no perfect or best way to introduce a young queen into a colony. I prefer using the mailing cage the queen is received in (if she is shipped from the South or California) or, if she is a queen we have reared, placing her in a cage with some queen cage candy and allowing the bees to release her by consuming the candy.

Our studies on queen recognition by workers have shown that bees taken from a colony and given a new queen will come to recognize her as their own after about 24 hours. There is no reason to delay a queen's release beyond that time, but it is important that she not be released early. If a queen is released too early, before she is accepted by the bees as their own queen, she will be balled and possibly killed.

If the parent colony's queen is, by mistake, among the workers used to make up a nucleus colony with a new queen, the bees will not accept the new queen as their own even after many days. When their own queen is present, bees have a remarkable ability to recognize her and reject all others, whatever the circumstances and ages of the two queens.

It is often possible to exchange laying, mated queens of the same age and physical condition between two colonies. The chances of successfully doing so are increased if a honey flow is in progress. It is also true that the smaller the number of bees in the nucleus colony, the greater the likelihood that the introduction of a new queen will be successful. Because all this is true, beekeepers have found that seemingly strange methods of queen introduction often may work.

One will find many different queen introduction techniques described in bee journals. These include dunking queens in honey

40. This is a mature colony being requeened using the mailing cage as an introducing cage. The cage is placed near the brood, between two frames and with the candy end up. The cork, which is placed in the hole of the candy end by the queen breeder, has been removed, exposing the candy. The candy will be eaten by worker bees in the new colony and the queen released.

41. *A queen cage, with the candy seen in the lower hole, is placed between two frames in a super and is about to receive a package of bees. The candy end of the cage is slightly higher than the opposite end; in the event that a worker bee dies, it will not fall against the candy and perhaps block the exit.*

42. This is a standard innercover used to separate a nucleus colony from a parent colony, which will be below the innercover while the nucleus colony is above. Pieces of 8-mesh hardware cloth have been nailed above and below the innercover hole. A piece of the edge of the innercover (lower right) has been cut away to make an entrance; in this way the nucleus has an entrance on a side of the hive different from that used by the parent colony.

before setting them free in a colony, driving all bees out of the hive with a repellent and placing the queen in the center of the confused mass when the bees reenter the hive, and feeding the bees in the hive and the queen some scented sugar syrup so that their odors will be the same. Some beekeepers prefer confining the queen in a cage over a patch of comb where she can lay eggs and still be fed from the outside by workers. The queen must be released after a day or two. All of these methods will work sometimes, but none is perfect. I do not recommend them and suggest that the time-tested use of a queen cage is best, although it is not infallible. The queen cage must be removed, but that can be done at the next routine visit to the apiary.

When we make nucleus colonies we usually use one frame with a small amount of brood, bees to cover (which usually means the bees shaken from two or three other frames), and one frame full of honey with some pollen. The queen cage is placed between two frames, close to the brood, with the screen fully exposed and the candy end up[1] (with the cork removed from the candy end, of course). It is not necessary to remove the worker bees sent with the queen. On the other hand, if one is caging a queen for introduction into a nuc, it is not necessary to add workers to the queen cage if there is a delay of only five to 10 minutes before the queen is placed in a colony.

THE BEST METHOD OF REQUEENING

Requeening large colonies is difficult. Those who do so routinely have a variety of techniques. I believe the best of these is to raise a frame with some brood, bees to cover and a frame of honey into a super above some kind of double-screened board, a board with two screens about one-quarter inch apart, and place both above the parent colony. A young mated queen, or a queen cell, is introduced into the new unit which has its own entrance.

This method is sometimes used as a swarm prevention technique as well. Removing one or two frames of brood from the center of a crowded brood nest and replacing them with empty combs may be all that is necessary to prevent the colony from becoming congested and swarming. The board separating the two colonies is therefore sometimes called a swarm board. We make these boards by placing two pieces of 8-mesh hardware cloth, one above and one below the hole in an innercover. The entrance is made by cutting away a piece in the innercover side, or by using a super with a hole drilled in it to give a colony an upper entrance in the winter. The use of two screens, separated by about one-quarter of an inch, keeps the bees in the two colonies apart. Heat from the larger parent colony passes into the upper colony and may be of assistance in keeping brood warm.

The new colony is usually made in June, the peak of our

---

[1] If the candy end of the queen cage faces down, worker bees that die within the queen cage may lodge against the candy, making it impossible for the queen to escape.

swarming season. When the young queen is established (has her own brood and young bees) in her new colony, it is ready to be combined with the parent colony after the latter is dequeened. This, as discussed, is usually done about August 1. Until that time the new colony may be kept above the parent hive or set off to one side. Whichever system is used often depends upon how much manipulation is necessary with the larger hive.

I prefer putting queen excluders on hives about July 7-10 in

43. A small nucleus colony (above) has been combined with a larger parent colony (below) using the newspaper technique. The newspaper, with three to five slits in it, will be chewed away by the worker bees; this will require several hours and by that time the colony odors will have mingled and there will be little or no fighting among the bees in the two units.

New York state. This is at the end of the swarming season, and driving the queen down into one super with a bee repellent and crowding the bees at this time of the year will not cause swarming.

Twenty-four days after the excluders are put into position, the supers of honey may be removed and there will be no brood in them. At the same time, the old queen who is confined to a single super may be found and killed. The excluder is removed from its position above the bottom super. A single sheet of newspaper, with several slits in it, is put in its place and the super with the young queen and her brood is placed above the newspaper. The queen excluder is now placed above the second hive body and the colony is supered for the fall honey flow.

It should not be necessary to inspect the colony again until fall, when the last inspection for disease is made and the honey is removed. In our experience the requeening success rate is 95 to 100 percent.

CHAPTER 9

STOCK SELECTION AND IMPROVEMENT

While attending shows where horses, cows, sheep and other animals are exhibited, I am always impressed by the fact that some farms win the majority of the first-place prizes year after year. These farms often sell some of their best stock, which would suggest that in time they would suffer severe competition from those animals and their offspring. However, this is rarely the case and the same farms continue to win. Admittedly, the better farms have good stock but they do not have a monopoly. Such farms have good management and husbandry. In addition to knowing how to select good stock, they take good care of their animals. They also may be good salespeople!

I am often asked where we buy our queens. I don't hesitate to say that we use several sources, but like most beekeepers I have a favorite queen breeder whom I know and whose judgement I respect. He is also a man who produces honey himself and is concerned with the production of good stock. He lives in an area where nectar and pollen plants abound and which has had a good queen-producing reputation for a number of years.

I have digressed in the above two paragraphs to emphasize that there are some underlying considerations when one selects breeding stock. One can be misled, by outward appearances and a good promotion program, into thinking that something (even a

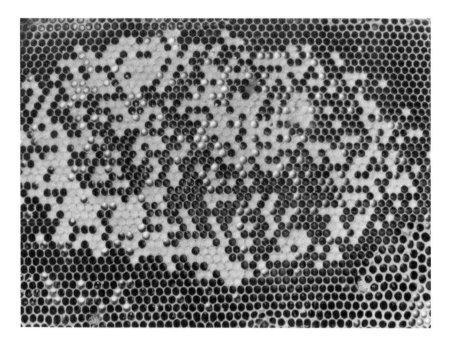

44. This is a poor brood pattern indicating a poor queen. This frame contains many cells with capped pupae, but they are not adjacent to one another. Mixed with the capped cells are cells containing larvae. Poor brood patterns may result from many things, including infertile eggs and disease, but most frequently it is the result of having an old queen which does not lay compactly.

book) is far superior to what it really is; but when one understands these considerations, it is possible to make sound selections.

What are the major considerations in selecting breeding stock? For most beekeepers, honey production is most important. For beekeepers who rent bees for pollination there are other considerations, although bees that are good honey producers also are good pollinators because they are more active in the field. However, high pollen-collecting bees, which are superior pollinators because they emphasize pollen collection, have been selected and may be used to enhance pollination.

Disease resistance is the second most important consideration. Chemicals are and will continue to be important in treating some diseases. However, stock that is resistant to diseases such as tracheal and varroa mites, chalkbrood, European foulbrood, and others has the greatest value.

Bees that are widely used today gather too much propolis in my opinion. These bees (Caucasians and Carniolans) are used because of their presumed resistance to tracheal mites. Propolis is naturally toxic to a wide variety of microbes, and perhaps it is necessary to protect these races of bees. However, propolis is also a nuisance and one's fingers are soon a sticky mess when examining colonies. Years ago there was less propolis in hives than there is now, and I like to select bees that use less propolis.

45. This is a good brood pattern. In the lower right is a small patch of drone brood, but this does no harm. On the two outer edges are patches of capped brood, probably close to emergence. In the center is a compact mass of young larvae, probably about three days old; moving left and right from the center, and between the three-day-old larvae and the capped brood, are larvae one and two days old, although the latter cannot be seen clearly. When brood of like age is close to brood of the same age, one has a good brood pattern. Queens normally lay in concentric patterns like this.

46. *This is a beekeeper checking his colony for the brood pattern. Even from this distant view one sees a compact mass of capped brood, indicating a good queen.*

47. This is a compact brood pattern, but in a frame with considerable drone brood near the bottom. Some beekeepers remove combs such as this from their brood nests; however, current data indicates that having a large number of drones present does not decrease honey production.

48. This is an excellent brood frame with mature pupae. The queen has failed to lay in only a few cells and obviously is young and vigorous.

49. This is a mediocre brood frame which nevertheless shows the concentric patterns in which a queen lays eggs.

There are other considerations such as gentleness, ability to overwinter, and color. Certainly light colored queens are easier to find than dark queens are.

In northern states the least amount of brood is reared in October and November. I thought for a long time that this was controlled by day length—that is, less brood rearing takes place on shorter days. However, on the southern tip of Florida there is usually an autumn honey flow from Brazilian pepper and melaleuca. This stimulates brood rearing in the fall, and it has been reported that in some years the bees may swarm. It is important to know what natural variations in brood production exist in one's area so as to make sound decisions in selecting stock.

SELECTING QUEENS FOR BREEDING STOCK

Calderone and Fondrik showed conclusively that one can increase honey production by using queens from the most productive colonies as breeding stock. It is not necessary to weigh colonies all year to determine which are best; short-term weight gain, from mid-spring through the end of the honey production period, is enough.[1]

This system of selecting breeding stock works even when the queens mate freely with whatever stock is available. Of course, it is helpful if one encourages high-producing colonies to produce more drones. This may mean moving low-producing colonies away from the mating area so that their drones are not present.

I suggest that, in addition to honey production, one should check a queen's brood pattern, as this is indicative of her physical condition. A breeder queen should be in good health when she is used as a breeder and also during the honey-producing season when she is judged. A good queen has a compact brood nest. In colonies with good brood patterns, eggs are adjacent to eggs, larvae to larvae, and pupae to pupae. In other words, adjacent brood should be approximately the same age. Queens lay eggs in ever expanding concentric circles. A queen should lay in cells as soon as they are cleaned by house bees and before honey and pollen are stored in them. There is some competition in this regard; worker bees will place both pollen and new honey in the center of the brood nest if given an opportunity. When one finds an overabundance of pollen and honey in the brood nest, there is usually something wrong with the queen.

It is best to use more than one queen as a breeder. There is always a small danger in using a single queen because there may be a hidden undesirable quality. Of course, if one is growing only a few queens this is not possible. The same is true in encouraging drone stock. Adding drone combs to several good colonies will encourage the production of drones.

I suggest that making notes about other qualities is a good idea as one examines colonies throughout the year. We all consider different traits to be important. Making notes about these will aid in the final selection process.

---

[1] Calderone, N. W. and M. K. Fondrik. Selection for high and low, colony weight gain in the honey bee, *Apis mellifera*, using selected queens and random males. *Apidologie* 22: 49-60. 1991.

Tested queens

A variety of names have been given to queens kept under observation for the purpose of studying their colonies, productivity, resistance to disease, etc. Most commonly these have been called tested queens or select queens. They have been sold under these names to other beekeepers with the idea that they would be used as breeder queens.

As discussed, the fact that queens mate many times, each time with a different drone, and that the sperm does not thoroughly mix in the spermatheca means that the offspring produced at one time may be quite different from that produced a few months later. One rarely sees the words "tested queen" used today because most breeders now understand that it is a meaningless term. Only persons who grow queens in isolation and have complete control over their drone stock are in a position to produce tested queens. Even under these circumstances, variations in the offspring may occur.

CHAPTER 10

CONTROL OF NATURAL MATING

Natural mating in honey bees was discussed in Chapter 2. We are aware that bees today are much the same as they were thousands of years ago because of our inability to control mating. We have moved bees from one area to another so that the races are not pure, but still we have not made any great strides in stock improvement.

I think that only a few beekeepers, usually ones with many hundreds of hives, have selected colonies that have many of the qualities we seek in bees. A few years ago, for example, when we were collecting venom, we worked with a beekeeper who had made a strong effort to eliminate the more aggressive bees from his 2,000+ colonies. We soon found that the quantity of venom we were collecting from these bees was much smaller than normal. We began collecting venom from apiaries belonging to another beekeeper and observed that the quantity of venom we collected increased; still, as far as we could observe, the colonies belonging to both men were of equal strength. This is a subjective—not objective—observation, but I think it is correct that the first beekeeper owned bees that were less inclined to sting and that he had been successful in selecting these bees.

Mating in confinement

No one has ever been able to mate queens in a cage successfully, although many people have tried to do so. Dr. Jan Nowakowski and I reviewed the subject in the June, 1971 issue of *Gleanings in Bee Culture*. One of my students, Dr. John Harbo, prepared an annotated bibliography on the attempts that had been made.[1] Dr. M. V. Smith of Guelph University reported he had a single queen mate in confinement (in an indoor flight room), but he was not able to repeat the observation.[2]

A great variety of cages have been tested to force bees to mate in confinement. We tested a design that sounded reasonable; it involved building a cage covered with cloth so that the light level was reduced. All of the cages designed for mating in confinement were built before the discovery of drone congregation areas. I often have thought that a very large cage built over a congregation area might be worthwhile, but I never have had the courage to spend the amount of money needed to do so.

I have no doubt that many people in the future will make attempts at mating queens in confinement. I am reluctant to discourage them because I know that some little quirk in a method often is all that is needed for success. Still, there is not much to encourage one to enter such a venture. There is a very slight possibility that African or Africanized bees might behave differently from European bees in this regard.

Mating in isolation

Few queen breeders have taken advantage of the opportunity offered by mating queens on islands or in mountainous areas. Using the isolation offered by mountains is probably more difficult because most are poor foraging areas for bees and feeding is required. Our knowledge of natural mating by honey bees has increased greatly in the past few decades. The information we have suggests that improvement of our present conditions could

---

[1] International Bee Research Association, Bibliography No. 12. Available for a small charge by writing IBRA, 18 North Road, Cardiff, Wales, CF1 3DY, U.K.
[2] Smith reported on this in the July, 1961 issue of *Bee World*.

50. This is a cage built in Africa in the 1960s in which some matings by queens and drones were claimed. The height may be judged by the door, which is about six feet high. It is not known if African bees use congregation areas for mating, as European bees do. Hundreds of cages have been built by people hoping they might confine queens and drones and obtain natural, pure mating. All such attempts have failed. Photo by P. Papadopoulo.

be made. There are many problems with our present system which detract from progress in stock improvement.

Our present system has the following faults: one piece of data we have, which admittedly is only circumstantial, suggests that drones fly farther than queens do when searching for mates; still, most queen breeders keep their drone-producing colonies in or very near their mating yard. Hobby beekeepers are everywhere, and because many of them give their colonies little management and tend to heavily weigh such factors as gentleness, their drones may negate at least a portion of what a queen breeder attempts to accomplish. Wild colonies in trees and buildings also may be a harmful factor; on small islands the beekeeper may locate and destroy them. A very small number of drones may fly many miles, perhaps as many as six or eight, to mate; thus it only would be possible to practice close control over the mating process on relatively small islands.

I am aware that one successful queen breeder isolates some of his stock in a mountainous region, where keeping four to five colonies would be a marginal operation. Persons interested in studying and taking disease-resistant stock could take advantage of such areas.

Still, from a practical point of view, our present system is working and supplying our needs. My thought is that things might be better. I am aware that many queen breeders attempt to isolate themselves by surrounding their mating yards with apiaries of their own stock. I've also heard of the occasional beekeeper who routinely requeens colonies belonging to nearby hobbyist bee-keepers.

INSTRUMENTAL INSEMINATION

It is possible to instrumentally inseminate virgin queen honey bees. Attempts to do so were made in the early part of this century, and became practical as a result of studies published in 1927 by Dr. Lloyd R. Watson while he was a graduate student at Cornell University. Several improvements in the insemination equipment have been made by Drs. Harry H. Laidlaw, Jr. of the

University of California and John R. Harbo of the U. S. Department of Agriculture. Instrumental insemination is a delicate operation and attention must be paid to detail.

Even under ideal circumstances, it is difficult to force a full complement of sperm into a queen's spermatheca. Instrumental insemination is an important research tool and has been critical in a number of experiments. However, as a practical method in commercial beekeeping its use is limited.

CHAPTER 11

REARING QUEENS AND MANAGING AFRICANIZED BEES

Honey bees from Africa were taken to Brazil in 1956 for the purpose of improving honey production in that country. The bees used in South America until that time were the same European races that had been imported into Central and North America. No honey bees are native to the Americas. However, beekeeping was never very extensive in the northern, tropical part of Brazil because European bees did not thrive there. It was hoped that the African bees, being a tropical race, might improve the situation. Although the African bees are more aggressive, they are the same species; thus it was presumed that aggressive characteristics could be bred out of the Africans, or that they could be crossed with less aggressive bees to produce gentler offspring.

During the first years of Africanization in Brazil, many bee-keepers—especially hobbyists—gave up beekeeping because of the aggressiveness of the new bees. However, despite an initial downfall, honey production in Brazil has increased since the introduction of the African bees. Beekeepers soon learned that with modifications to their management programs, the new bees were excellent producers. It is interesting to note that there is a well-established, American-type beekeeping industry in parts of Africa using these same bees. Several African beekeepers have

51. Examining a natural swarm found in Kenya, east Africa; the bees in this swarm are fully engorged and gentle, as is usual in a swarm of European honey bees.

visited and/or worked in the United States to learn our techniques.

THE TERM AFRICANIZED

The common honey bees in Brazil today are called Africanized. The term is used to differentiate them from bees in Africa and the native stingless bees in Brazil. They are not a new race or species of bee. They are bees that have been crossed with the original European bees. They are predominantly African. Their size and behavior is similar to that of the bees on their native continent. On average, Africanized honey bees are about ten percent smaller than their European counterparts, as are African honey bees. It is an amazing fact that the importation and escape of a small number of queen honey bees from Africa could lead to over three or four million colonies of Africanized (nearly African) bees in South, Central, and North America today.

52. *Examining a colony of bees in a plastic hive in Brazil; as seen here, the Africanized bees have a tendency to run onto the top bars and over the side of the hive. European bees will do this only rarely.*

AGGRESSIVENESS IN AFRICANIZED BEES

Studies in Brazil, financed by the United States Department of Agriculture, showed that Africanized bees are more aggressive in warmer climates. These studies compared the aggressiveness of colonies headed by the same queens in the tropical and sub-tropical parts of Brazil; unfortunately, similar studies were not conducted in the temperate parts of South America. Africanized bees are present in Argentina (south of Brazil), where some people have reported a great number of aggressive colonies. Argentina has been one of the world's major honey-exporting countries for many years. It is important to note that Argentine honey production has not been adversely affected by African bees, and Argentine beekeepers do not appear to be concerned about them today.

Methods of measuring aggressiveness in Africanized bees were first devised by Dr. Antonio Stort, a Brazilian who wrote a thesis on the subject. His methods have been copied by several North Americans who have since made observations of these bees. Stort found that black balls of felt or suede could be dangled in front of colony entrances for a given period of time; one could then count the number of stings in the ball and thereby obtain an estimate of aggressiveness. He also discovered that bees of different temperament varied in the distance over which they would pursue a presumed enemy; this, too, gave a measure of aggressiveness.

A little-studied technique reported from east Africa is keeping colonies of aggressive bees in bee houses where the hives are protected from the hot rays of the sun. The report states that working with colonies in a bee house, as is done in parts of Europe, is less difficult because the bees are gentler. On the other hand, bee houses often are filled with smoke as a result of excessively smoking the colonies and may be uncomfortable for that reason. Nevertheless, this technique should be studied in the United States to determine if it would be helpful, especially in the warmer states.

Queen rearing with Africanized bees

During the past 20 years I have visited Brazil many times and spent a total of about six months there, especially in the state of São Paulo where the African bees were first introduced. I spent most of my time at the University of São Paulo's genetics laboratory, located in Ribeirao Preto which is a city of about 400,000 people. The head of the laboratory for many years was Professor Warwick E. Kerr, who brought the African bees into Brazil; one of his students is now in charge of the laboratory. The first introduction of African bees was made near this laboratory, and so it is in Ribeirao Preto that we find beekeepers and researchers with the most experience in working with Africanized bees. São Paulo state is agriculturally rich and for many years was the chief coffee-growing region in Brazil. Citrus is an important crop in the state today, and as is true everywhere in the world with a strong citrus industry, there is also a strong beekeeping industry since citrus is a good producer of nectar.

53. This is an apiary in Brazil. Each colony is on its own hive stand so that any vibrations made as a result of examining one colony will not arouse the bees in another hive.

HOW QUEEN REARING TECHNIQUES VARY WITH AFRICANIZED BEES

The laboratory in Ribeirao Preto has been growing queens almost since the introduction of the African bees. The queen breeder has been in residence for over 25 years. He grows Africanized and European honey bee queens and their hybrids side by side for experimental purposes, and thus he has experience with all three. European stock is constantly being imported for breeding purposes.

There are only a few basic differences in growing Africanized queens. These include using larger mating nucs, taking care where one locates colonies, and dressing properly for work in the apiary. Brazilians have found that larger smokers than those used in North America are more practical. African queens mature in

54. This student in Brazil is holding the caged queen from an Africanized swarm in her hand, and the bees cluster around the queen. The engorged bees are gentle. Bees in a swarm, whatever their origin, will not abandon their queen and may be led or carried anywhere as long as they are kept in the sunlight. Manipulating bees in this manner is not difficult, but only should be done by experts in bee culture.

about 15 $1/2$ days instead of 16. Queen finding is more difficult in the case of Africanized bees because the queens are smaller and darker. They run more on the combs along with the workers. Much of the rest is routine for all races of bees.

Many queen breeders and royal jelly producers in Brazil use hybrid bees. These are colonies with Italian queens that are openly mated with Africanized drones. These bees are easier to work with, but still are quite productive compared to Africanized bees.

ABSCONDING BY AFRICANIZED BEES

There is a tendency for the bees in small Africanized bee mating nucs to abscond at the same time a virgin queen takes her mating flights. This sometimes occurs with European races but is not regarded as a serious problem with European stock. As a result, queen breeders in Brazil use larger mating nucs, usually three standard, full-depth frames and at least 3,000 bees; this is at least twice the size of the mating nucs used by most queen breeders in the United States today. Small nucleus colonies of Africanized bees may be used if several square inches of brood are present. However, keeping brood in small mating nucs, especially on a large scale, is difficult.

For many years several beekeepers in South Africa placed bait hives in areas where they knew a number of feral colonies lived, thus potentially attracting a large number of swarms. This appeared to be an easy way of obtaining a large number of colonies for honey production with little effort and cost. However, after several years of experience it was found that swarms obtained in this manner had a greater tendency to swarm and/or abscond; that is, they were genetically different. If colonies are obtained in this manner today they may be requeened.

As for queen rearing, the use of strong cell-starting and cell-building colonies is no different than it is with European bees. In brief, absconding is not a problem with large, well-fed Africanized colonies.

Requeening

We now have many years of experience in requeening and exchanging queens between Africanized and European colonies in Brazil. It is never easy to requeen large colonies of European bees and the same is true of large colonies of Africanized bees. However, we have found that we can requeen and exchange queens between these races as easily as we can within each race. In both cases the smaller the colony, the easier it is to requeen. Combining nucleus (small) colonies with new queens and mature colonies that have been dequeened is favored over other techniques. The queenright unit is placed on top of the queenless one so that bees from the queenless colony do not come quickly into contact with the new queen. A piece of newspaper is placed between the two units. The paper is slowly chewed away by the bees. Small colonies are easily requeened using normal queen cages that are placed in the brood nest area with the candy end up. The bees eat the candy away and release the queen.

Dressing for the Africanized apiary

Beekeepers who work with Africanized bees dress carefully. They wear good boots—usually smooth finished, light-colored leather or white rubber boots. They wear light-colored pants and shirts or white coveralls. It is useful either to have elastic bands at the bottom of the trousers and arms of the protective clothing, or to tie these in place so that bees cannot crawl underneath. The faces of the veils are made with wire, not cloth. An important feature of the veil is that the wire is painted black on the inside but white on the outside. One cannot see well through a wire cloth that is any color other than black. Dark colored clothing and black boots and veils are repugnant to honey bees. Bees in an aroused apiary will hit at a black wire veil much more often than a white one.

To make a proper veil, one starts with wire that is enameled black but has white or aluminum paint applied on the outside. One must use only a light coat of white paint so as to not plug the wires. The paint may be applied with a spray can, roller, or brush.

55. *A watering trough for bees in tropical Brazil. Water lilies grow in the water. The bees walk on the lily pads when drinking. I saw no bees drown in this pool.*

The hat for the veil should be made of light colored straw because it is cooler and the color does not upset the bees. Beekeepers in Brazil do not care for the heavy leather gloves we frequently use in the United States; they are too hot to wear and retain sweat odor too long. New leather gloves also have an oil or leather odor which is offensive to the bees. Rubber gloves for washing dishes are used by many beekeepers, if they use gloves at all.

Location of apiaries

The university laboratory where I work in Brazil has an apiary of over 100 colonies, including both large, populous colonies and small ones. These are immediately behind the laboratory where at least 20 people work. Two other laboratories are nearby and within 300 feet of the apiary. All of the people in these laboratories are aware of the bees and respectful of them. They do nothing to antagonize the bees. Likewise, and equally important, those who work with the bees are careful with the use of smoke and the way they manipulate the colonies. One does not walk up to a colony and open it without first applying smoke in the proper manner.

In general, beekeepers in both South Africa and Brazil keep their apiaries some distance—at least 500 to 1,000 feet—from homes and barns. It is best if the apiary is surrounded by woods and a fence so that animals will not mistakenly wander into it. In a few cases cows, horses, chickens, etc. have wandered into apiaries in the United States and been killed by angry bees. Such incidents appear to occur more often with Africanized bees, although relatively few have been carefully documented in either case.

I have observed many colonies of bees living in buildings, boxes, crates, and hollow trees in Brazilian cities, apparently in harmony with people. The discovery of feral colonies in such places in Brazil is much the same as it is in the United States. As long as food is available for the bees, they will take advantage of it and live in the vicinity. The only way to eliminate or reduce the number of honey bees in a city is to prohibit the growing of nectar and pollen producing plants.

The proper use of smoke

As discussed earlier, smoke is the best defense against aggressive bees of all races anywhere in the world. We have no data, but we presume that smoke fouls or deadens the sensory receptors on the antennae. There is no evidence to show that smoke from one type of fuel is any better or safer than smoke from another. Smoke that contains anesthetics should not be used; these have been tested and there is great danger of honey bees suffocating when large numbers are anesthetized. Those who work with Africanized bees make certain that their smokers are lit and ready for use before entering the apiary.

CHAPTER 12

MISCELLANEOUS CONSIDERATIONS

Growing queens requires a broad knowledge of beekeeping. Some important questions not touched upon in the main portion of this text are reviewed here. This by no means includes all of those extra subject areas a queen breeder must become familiar with, but only some of those that are especially important.

ROBBING

Robbing bees are fascinating to watch and study, but a horrible nuisance for one who has work to be done in the apiary where it is taking place. Once robbing starts it is almost impossible to stop it on the same day; many beekeepers have given up in an apiary where they have made a mistake and caused it to start. Beekeeping operations should be geared to prevent robbing. In large part this involves understanding when robbing may occur.

It is an interesting fact that honey bees prefer natural sources of food, nectar, and pollen over unnatural ones such as sugar syrup, honey, and other sweets that may be exposed and available to them.[1] So long as pollen and nectar are available, honey bees will not explore or accept other food. Bees have a threshold of perception and acceptance. The first term concerns what they

---

[1]It is interesting to think about the evolutionary basis for this fact. Solitary and semisocial bees, such as the leafcutters, carpenter, and bumblebees, will not feed on any unnatural source of food, but will only take natural food. Many wasps, both

may distinguish, and the second, what they will accept as food. The threshold of acceptance varies from day to day depending upon what is available.

A practical beekeeper must be aware of what his bees are feeding on and when no food is available. When robbing is likely to occur, he opens and closes hives quickly, leaves no honey exposed, and makes certain no burr comb with honey is left on the ground. When robbing may be especially serious, one may use a cage, usually about three by six feet and five or six feet high, without a top, which one may get inside of and carry. The cage needs to be big enough to cover one hive and the beekeeper while essential operations are made.

DISEASES OF BEES

Honey bees, like most animals, suffer from a wide range of pests, predators, and diseases.[2] However, the picture has become more complicated because of the introduction of three new diseases into North America in recent years: chalkbrood was found in California around 1968; tracheal mites were discovered in Texas in 1984 (but had been found in Mexico a few years earlier); and varroa mites were found in Wisconsin in 1987. By the time each of these new problems was discovered it was already widespread, and eradication and the prevention of spread across the nation was not possible.

We have a good chemical control for varroa mites, a fair chemical for controlling tracheal mites, and nothing that is effective against chalkbrood at present. I estimate that during the first six or so years after they were found in Texas, well over half of the colonies in the United States died because of tracheal mites. Commercial beekeepers have been able to recoup these losses

---

solitary and social, will feed on both natural and unnatural foods; for this reason they are often pests around honey houses, garbage cans, picnic benches, etc. In fact, a small number of species apparently can substitute our foods, such as hamburgers, hot dogs, soda, ice cream, and beer, for the insects they normally capture, eat, and feed to their young.

[2] *Pests, Predators and Diseases of Honey Bees*, edited by Roger A. Morse and Richard Nowagrodzki, Second Edition 1990. Cornell University Press, Ithaca, NY 14850.

rapidly. I estimate that we have lost five percent or more of our honey crop during each of the past 20 or more years because of chalkbrood. Varroa mites have caused the deaths of thousands of colonies, especially in those states where migratory beekeeping is practiced on a large scale. Chemical control is necessary for varroa mites or the colonies will perish.

In the cases of chalkbrood and tracheal mites, we believe the most susceptible stocks already have died and that most bees today have some resistance to both. Bees that are resistant to varroa mites have been found both in Europe and the United States, but they are not common. African and Africanized bees appear to be naturally resistant to varroa mites.

As agriculture has changed (intensified) in recent years and government budgets have grown leaner, we have seen some states abandon their apiary inspection programs while in others the programs have been greatly reduced. This means that bee-keepers must be increasingly responsible for the diagnosis and treatment of their diseased colonies.

Diseases that may be diagnosed in the apiary by the bee-keeper include chalkbrood (white, grey, and black mummified larvae), varroa mites (pinhead size mites on brood, especially drone brood), American foulbrood (dead larvae lying flat in the bottom of a cell), European foulbrood (dead larvae lying curled in cells), and sacbrood (dead larvae in a greyish, watery sac). Nosema usually may be diagnosed by examining the guts of worker bees; these are usually white and swollen in infected bees. Tracheal mites can be found in the large breathing tubes (prothoracic tracheae) in the thorax, but only with a microscope. Many bee-keepers are buying microscopes in self defense.

We have come to understand a great deal about disease control in honey bees so that a few guidelines can be offered: (1) Keep colonies in the sunlight where they will be warmer and drier. (2) Place apiaries on land that slopes to the east or south for maximum exposure to the sun. (3) As much as possible, face colony entrances east and south. (4) Use some type of hive stand or wooden or stone structure to raise colonies a few inches off the ground so that they stay dry. (5) Tilt colonies slightly forward so that bottomboards do not collect rain and stay wet. (6) Place bees near fresh, clean water. The above guidelines are especially important in colonies that are being used to rear queens. As far as we know, queen honey bees may suffer from the same diseases which affect workers. Understanding much of the above advice is making it possible for colonies to remain dry and warm and hold a brood rearing temperature with ease.

Routine annual chemical treatments for nosema, American foulbrood, European foulbrood, and varroa mites should be practiced in areas where these diseases are common. In some areas where there is little migratory beekeeping it may be possible to control American foulbrood by other methods, but only if one's neighbors are cooperative.

Beekeepers should support and encourage private and public programs to develop bees which are resistant to, or at least can tolerate, several of the common diseases. Those who grow queens for sale or for their own use can do much to aid in this regard by selecting breeding stock from disease-free colonies. It is likely that if one uses high honey yields as a criterion for stock selection, bees in such colonies will be more disease resistant than common stock. It is also possible that in the not too distant future, beekeepers also may look carefully at feral colonies (wild colonies in trees and buildings) for stock that may have survived unattended because of a natural resistance to one or more diseases.

BEE PESTS AND PREDATORS

There are a number of bee pests and predators which cause special problems for queen breeders. Most notable among these are birds, robber flies, and especially dragonflies. All of these are general feeders, that is, animals that feed on a wide variety of other animals. Queens are probably attractive to these predators because of their large size.

A Florida east coast queen breeder once told me that there were two weeks during the spring when a particular species of dragonfly was so abundant in his area that it was impossible for him to grow queens because of the high predation rate. Another Florida beekeeper told me about being in an apiary where there were so many pieces of bees falling and hitting him that it reminded him of rain. Dragonflies often feed while flying and usually consume part of the thorax only, allowing other parts to fall to the ground.

A small wingless fly, *Braula coeca*, is sometimes found in hives, and queens are particularly attractive to it. Over a hundred may sometimes congregate on a single queen. The flies apparently do not feed on the queen, but use bees to transport them around the hive or from one hive to another. When they are present in large numbers, it is believed that they may hinder the queen's movement and egg laying.

In each instance where queens have special problems, it is usually localized. The best source of information is a local beekeeper. Probably no one of these special problems is as serious as to inhibit queen rearing in any place where bees are normally kept, except in the case of dragonflies, which may cause temporary disruptions.

ROYAL JELLY PRODUCTION

Royal jelly is a glandular secretion produced by worker bees and fed to young worker larvae, queen larvae throughout their larval lives, and adult queen bees. It is a thick, white fluid with a consistency reminiscent of heavy cream. Royal jelly is the food which makes queens different from workers, although precisely how this is done is unclear. Since queen honey bees live much longer than workers do, many people have ascribed magical, or at least special, powers to royal jelly. In many countries it is, or has been, sold as a human dietary supplement with claims that it will increase longevity. While royal jelly is excellent food for bees and, in fact, they could not survive without it, there are no data to substantiate the claims made for it as a human food. It is in demand by persons searching for an elixir. As long as the demand continues, I have no doubt that some beekeepers will devote some of their effort to its production.

123

Royal jelly is produced by growing queens in the normal manner. Collection of jelly happens on the third day of larval life while the developing queen is still in the second instar, or stage. At this time larvae are small and the supply of royal jelly in the cells is most abundant. The larva is lifted from its bed of jelly with a small spoon and discarded. The jelly is spooned from the cell or picked up with some sort of vacuum device.

Several papers have been written on how to produce royal jelly. Persons interested in this aspect of queen production could most profitably search old bee journals for methods advocated by the royal jelly producers. Little royal jelly is produced in the United States and Canada because labor costs are lower elsewhere and our beekeepers cannot compete with those in other countries.

Professor M. V. Smith of Guelph University, Guelph, Ontario wrote a clear and concise mimeograph entitled *The Production of Royal Jelly* in May, 1959. While this was not a formally published paper, it is available from some libraries and persons who have good reprint collections. Smith advocated the use of queenright colonies for jelly production. He found that 120 cells would produce one ounce of jelly if the harvesting was done on the third day.

FURTHER READING

One friend, who never kept more than 500 colonies, told me that one advantage of long, cold winter nights in the North was that they gave him an opportunity to read and reread his books and journals to determine how he might improve his management another year. This was a beekeeper who insisted on maximum production from each colony. It is true there is hardly anything new in the way of apiary practices. However, there are always refinements and improvements one might make and thereby save time and increase production.

There have been several books[3] published on queen rearing,

---

[3]Drs. T. S. K. and M. P. Johansson prepared a thorough and well-documented 26-page paper entitled "Methods for rearing queens" in the fourth issue of *Bee World* in 1973. Reprints are available from the International Bee Research Association, 18 North Road, Cardiff, Wales, CF1 3DY, U.K.

and several are out of print. Each book approaches the subject a little differently and has special methods which may be applicable in certain areas of the country. The following notes introduce the important authors of valuable queen rearing books.

Doolittle, G. M. 1888. *Scientific Queen-rearing.* **American Bee Journal. Hamilton, Illinois. 126 pages.**

This was a popular book and is often available through secondhand book dealers. Doolittle begins by saying he was introduced to beekeeping when he was seven years old. As far as I can determine, keeping bees was his sole means of earning a livelihood and certainly he was expert at it. He wrote a column for *Gleanings in Bee Culture* for 18 years and was considered one of the best of the early commercial beekeepers. Doolittle's second book, *A Year's Work in an Out-Apiary* was, together with Dr. C. C. Miller's *Fifty Years Among the Bees,* a bible of bee management. Many beekeepers still revere both books; I agree they are excellent and should be part of every beekeeper's library.

Kelley, Walter T. Undated. *How to Grow Queens for 15¢ Each.* **Walter T. Kelley Co., Paducah, Kentucky, Circa 1930. 20 pages.** *How to Grow Queens.* **Revised Edition, Circa 1940. 20 pages.**

Kelley wastes few words in his 20 pages; his advise is sound and his directions clear. He uses the following guidelines in selecting breeder queens: productivity, temperament, and color. While he says color is least important, it is clear he wanted to produce Italian (yellow) queens only. This short bulletin on queen rearing is now reprinted in the thirteenth edition of W. T. Kelley's *How to keep bees and make honey.*

Laidlaw, Harry H., Jr. 1977. *Instrumental Insemination of Honey Bee Queens.* **Dadant and Sons, Hamilton, Illinois. 144 pages.**

This is a pictorial manual and deals with instrumental insemination only. The book contains no information on queen rearing but will be of interest to those who do so. The book shows both Laidlaw- and Mackinson-style devices.

Laidlaw, Harry H., Jr. 1979. *Contemporary Queen Rearing.* **Dadant and Sons, Hamilton, Illinois. 199 pages.**

This is a well-documented book with clear and precise information on how to grow queens. Laidlaw, an emeritus professor from the University of California, began beekeeping as a teenager. He did much to improve instrumental insemination techniques for queen honey bees during his career. He is well-known among commercial queen breeders for his deep knowledge of the subject and brings to this book much information from his travels in the United States and abroad.

Pellett, Frank C. 1918. *Practical Queen Rearing.* **American Bee Journal. Hamilton, Illinois. 101 pages.**

Pellett wrote many books on bees. His queen rearing book is often available on the secondhand book market. Pellett's book is better illustrated that most. Included are several pictures of queen cells built under various conditions.

Snelgrove, L. E. 1949. *Queen Rearing.* **I. Snelgrove, Bleadon, Somerset. 344 pages.**

Many books have been printed in other countries on queen rearing. I cite only this one because it is probably the best known. Snelgrove wrote many books on bees. The type is large, making the text easy to read, and the book is moderately well-illustrated. Snelgrove was prolific, perhaps wordy; he devoted one 205 page book, which went through at least three editions, to the subject of *The Introduction of Queen Bees.*

INDEX

About This Book

    This book was set in Utopia and formatted in PageMaker 5.0 and composed from Macintosh files. New photographs were integrated into the book from Kodak Photo CD scans of slides. All photos are by the author unless otherwise noted.